Fueling Our Future: An Introduction to Sustainable Energy

One of the most important issues facing humanity today is the prospect of global climate change, brought about primarily by our prolific energy use and heavy dependence on fossil fuels.

Fueling Our Future: An Introduction to Sustainable Energy provides a concise overview of current energy demand and supply patterns. It then presents a balanced view of how our reliance on fossil fuels can be changed over time so that we move to a much more sustainable energy system in the near future.

Written in a non-technical and accessible style, the book will appeal to a wide range of readers both with and without scientific backgrounds.

Robert Evans is Methanex Professor of Clean Energy Research and founding Director of the Clean Energy Research Center in the Faculty of Applied Science at the University of British Columbia, Vancouver. He was previously Head of the Department of Mechanical Engineering and Associate Dean of Applied Science at UBC. He is a Fellow of the Canadian Academy of Engineering, the UK Institution of Mechanical Engineers, and the US Society of Automotive Engineers. Prior to spending the last 25 years in academia he worked in the UK Central Electricity Research Laboratory, for the British Columbia Energy Commission, and the British Columbia Ministry of Energy, Mines and Petroleum Resources. He is the author or coauthor of over 140 publications, and holds four US patents.

Fueling Our Future

An Introduction to Sustainable Energy

ROBERT L. EVANS
Director, Clean Energy Research Center
The University of British Columbia

CAMBRIDGE
UNIVERSITY PRESS

CAMBRIDGE UNIVERSITY PRESS
Cambridge, New York, Melbourne, Madrid, Cape Town, Singapore, São Paulo, Delhi

Cambridge University Press
The Edinburgh Building, Cambridge CB2 8RU, UK

Published in the United States of America by Cambridge University Press, New York

www.cambridge.org
Information on this title: www.cambridge.org/9780521684484

First published 2007

Reprinted 2008

Printed in the United Kingdom at the University Press, Cambridge

A catalog record for this publication is available from the British Library

ISBN 978-0-521-86563-0 hardback
ISBN 978-0-521-68448-4 paperback

Contents

Preface

Energy use, and its impact on the environment, is one of the most important technical, social, and public-policy issues that face mankind today. There is a great deal of research, and many publications, which address these issues, some of which paint a very pessimistic picture for future generations, while others point to a bright future through the use of new technologies or the implementation of new policies. Although a lot of excellent work is being conducted, much of the research necessarily tends to be quite narrowly discipline-based. Solutions to the problems caused by current patterns of energy use therefore often appear to be somewhat piecemeal in nature, and it is difficult for decision-makers and energy consumers to see the "big picture" which is really needed to understand and design truly sustainable energy processes. This book takes a systems approach to energy use, so that the complete consequences of choosing a particular energy source, or energy conversion system, can be seen. The concept of the complete energy conversion chain, which is a simple but powerful tool for analyzing any energy consuming process, is introduced to link primary energy resources through to the ultimate end-use. Looking at the complete consequences of any proposed energy technology in this way enables the reader to see why some proposed solutions are more sustainable than others, and how the link between energy consumption and greenhouse gas emissions can be broken. This simple systems approach is essential to provide a global understanding of how we can begin the transition to a truly clean and sustainable energy future. The environmental consequences of energy consumption and current energy use patterns are then summarized, providing the necessary background needed to understand the extent and complexity of the problem. Subsequent chapters outline the current state-of-the-art in sustainable energy technology, including non-conventional fossil fuels, renewable energy sources, and nuclear power. The challenging problems of developing a more sustainable transportation energy system are addressed in some detail, with a particular focus on road vehicles. Finally, some projections are made about how a sustainable global energy balance might be achieved over the remainder of this century. It is hoped that this book will be a valuable and thought-provoking resource not only for energy practitioners and students, but also for decision-makers and the interested public at large.

Acknowledgments

Few books such as this can be written without the author drawing freely on the ideas and thoughts resulting from discussions over many years with a wide range of colleagues, friends, and students. This one is no exception, and although there are far too many such individuals to name here, I would particularly like to thank my colleagues in the Department of Mechanical Engineering at the University of British Columbia for many stimulating discussions and debates. I would also like to thank the Master and Fellows of Pembroke College, Cambridge, who graciously granted me the privilege of being a visiting scholar during the 2004–2005 academic year, during which time most of this text was written. The editorial staff at Cambridge University Press were a delight to work with, and I am grateful to Dr. Matt Lloyd, Ms. Lindsay Barnes, Ms. Dawn Preston and Ms. Lesley Bennun for keeping me on track, and on time! My family, June, Kate, Jonathan, and Peter, were constant in their love and encouragement, without which I would never have been able to complete this task. And, finally, I dedicate this work to my granddaughter, May, who *is* the future.

Glossary

Barrel:
Crude oil can be measured both in terms of mass (tonnes), or by volume (cubic meters, or barrels). One barrel (Bbl) is equivalent to 35 Imperial gallons, or 42 US gallons. One tonne of oil is equal to approximately 7.35 Bbls.

Efficiency:
The efficiency of any energy conversion system is defined as the ratio of the energy or work output of the system to the energy input to the system. "Thermal efficiency" is usually used to describe the performance of a "heat engine," in which thermal or chemical energy is used to produce work.

Energy:
Energy can be defined as the "capacity to do work," and many different units are used. Energy can be found in many different forms, including chemical energy, as contained in fossil fuels, and thermal energy which can be related to the work which can be done as a result of a temperature difference in a substance. Electrical energy is that form of energy in which a flow of electrons can be used to do work with an electric motor, or to provide heat from a resistor network.

The basic energy unit in the SI (Système International) system of units is the Joule (J), where 1 J equals the energy required to do 1 N-m (Newton-meter) of work. In the Imperial system of units, still used in many English-speaking countries (particularly the USA), the basic unit of work is the foot-pound (ft.-lb.), and the basic energy unit is the Btu (British thermal unit). The energy required to heat one pound of water by 1 degree Fahrenheit is 1 Btu. The "mechanical equivalent of heat" states that 778 ft.-lbs. of work is the equivalent of 1 Btu.

Conversion between the two systems of units can be facilitated by noting that 1 Btu is equivalent to 1055 J.

Since the Joule represents a very small quantity of energy, values are often quoted in terms of multiples of one thousand. For example:

1 kilojoule	1 kJ = 10^3 J
1 Megajoule	1 MJ = 10^6 J
1 Gigajoule	1 GJ = 10^9 J
1 Terajoule	1 TJ = 10^{12} J
1 Petajoule	1 PJ = 10^{15} J
1 Exajoule	1 EJ = 10^{18} J

In Imperial units, it is common to use "millions of Btus," where:

1 MMBtu = 10^6 Btu

Because fossil fuels, and in particular crude oil, represents such a large fraction of total energy use in industrialized countries, total energy use is also sometimes quoted in terms of "tonnes of oil equivalent," or "toe." In other words, all energy use is converted to the equivalent energy contained in a certain number of tonnes of crude oil. A useful conversion factor is:

1 toe = 41.87 GJ

For large quantities of energy use, multiples of one thousand are again used. For example:

1 Megatonne of oil equivalent	1 Mtoe = 10^6 toe
1 Gigatonne of oil equivalent	1 Gtoe = 10^9 toe

Electrical energy use is usually measured in terms of the electrical power operating for a given amount of time. For example, the basic unit of electrical energy used by electrical utilities is a power of one kW acting for one hour, or 1 kWh. Therefore:

1 kilowatt-hour	1 kWh = 10^3 W for 1 hour
1 Megawatt-hour	1 MWh = 10^6 W for 1 hour
1 Gigawatt-hour	1 GWh = 10^9 W for 1 hour

Power:

Power is defined as the "rate of doing work," or equivalently, the "rate of using energy." The basic unit of power in the SI system of units is the Watt (W), defined as the power produced when 1 Joule is used for 1 second, or 1 W = 1 J/s. Again, multiples of one thousand are used to measure larger power quantities. For example:

1 kilowatt	1 kW = 10^3 W
1 Megawatt	1 MW = 10^6 W
1 Gigawatt	1 GW = 10^9 W

Engineers who design and operate thermal power stations sometimes make the distinction between "electrical power," using the suffix "e," and thermal power, using the suffix "t." For example, a large coal-fired power station may generate 2000 MWe of electrical power, while consuming coal at the rate of 6000 MWt, resulting in a "thermal efficiency" of 33.3%.

A more comprehensive list of energy unit conversions is provided in Appendix 1.

Part I Setting the scene

I

Introduction

The provision of clean, and sustainable, energy supplies to satisfy our ever-growing needs is one of the most critical challenges facing mankind at the beginning of the twenty-first century. It is becoming increasingly clear that the traditional ways in which we have satisfied our large, and growing, appetite for energy to heat our homes, power our industries, and fuel our transportation systems, are no longer sustainable. That this is so is partly due to the increasing evidence that emissions from fossil fuel usage are resulting in global climate change, as well as being responsible for local air pollution. It is also due to the realization that we are rapidly depleting the world's stock of fossil fuels, and replacement resources are getting more and more difficult to find and produce. The problem is made even more acute by the huge and rapidly growing appetite for energy in the developing world, where many countries are experiencing extremely high economic growth rates, leading to equally high demands for new energy supplies. In China, for example, total energy demand has been growing at an annual average rate of 4% in recent years, while in India it has been growing at 6%, compared with just under 2% in the rest of the world.

Global climate change, in particular the prospect for global warming, has put the spotlight on our large appetite for fossil fuels. Although there is considerable debate on the extent of the problem, there is no doubt that the atmospheric concentration of CO_2, one of the key "greenhouse gases," is increasing quite rapidly, and that this is likely due to mankind's activities on earth, or "anthropogenic" causes. The utilization of any fossil fuel results in the production of large quantities of CO_2, and most scientific evidence points to this as the main cause of increasing concentration levels in the atmosphere, and of small, but important increases in global average temperatures. Studies by the United Nations Intergovernmental Panel on Climate Change

(IPCC) have shown that the atmospheric concentration of CO_2 has risen from a level of around 280 ppm (parts per million) in pre-industrial times to nearly 370 ppm today, with most of the increase occurring in the last 200 years. The average global temperature over this same period appears to have risen by about 1 °C, with most of this occurring in the last 100 years or so. Computer modeling of the atmosphere by IPCC scientists, using a range of scenarios for future energy use, have suggested that over the next 100 years the concentration of CO_2 in the atmosphere may increase to a level between 540 ppm and 970 ppm, with a resultant rise in the global average temperature at the low end of 1.4 °C to a level of 5.8 °C at the high end. While mankind may be able to adapt easily to the relatively small changes in the global climate which would result from the lower estimate of temperature rise, at the higher end there would likely be significant and widespread changes, including a significant rise in sea-level around the world due to melting of polar ice caps and expansion of the warmer water in the ocean. At the extreme end there would also likely be increased desertification, particularly in low-latitude regions, and an increase in the volatility of global weather patterns. Of course, the widespread use of fossil fuels also results in significant local effects, in the form of increased levels of air pollution, primarily in large urban areas and centers of industrial concentration where the emission of oxides of nitrogen, unburned hydrocarbons and carbon monoxide lead to "smog" formation. These localized effects can result in serious health effects, as well as reduced visibility for the local population.

When energy use in any economic sector is examined in detail, the end-use can always be traced back to one (or more) of only three primary sources of energy: fossil fuels, renewable energy, or nuclear power. In order to understand the full implication of changes to our present pattern of energy utilization, however, it is necessary to consider the effects of any proposed changes on the complete energy system from primary energy source through to the final end-use. This is sometimes referred to as a "well-to-wheels" approach, in a reference to the complete energy supply and end-use pattern associated with providing fossil-fuel energy to motor vehicles. The same kind of assessment can be used to study any energy system, however, by considering the "energy conversion chain," which links primary energy sources to energy "carriers" like refined petroleum products and electricity, through to its ultimate end-use in the industrial, commercial, residential, or transportation sectors. This approach, which is outlined in more detail in the next chapter, is used throughout the book to provide an analysis of all the steps required in converting a primary energy source into its final end-use form. In this way all of the energy losses, and pollutant emissions, inherent in each of the conversion steps are taken into account so that a complete assessment of the overall energy system may be obtained. The need to establish a more sustainable global energy supply, without the threat of irreversible climate change, or the health risks associated with local air pollution, has led to many suggestions for improving current energy use patterns. Often, however, solutions that are proposed to address only one aspect of the complete energy conversion chain do not address in a practical way the need to establish a truly sustainable energy production and utilization system. This, as we shall see in later chapters, appears to be true for the so-called "hydrogen economy" which promises to be "carbon-free" at the point of end-use, but may not be so attractive if the complete energy conversion chain is analyzed in detail from primary source to end-use. By analyzing the complete energy conversion chain for any proposed changes to current energy use patterns, we can more readily see the overall degree of "sustainability" that such changes might provide.

The growing global demand for energy in all of its forms is naturally putting pressure on the declining supplies of traditional fossil fuels, particularly crude oil and natural gas. The large multinational energy companies that search for, and produce, crude oil and natural gas report that greater effort (and greater expense) is required to maintain traditional "reserves to production" levels. These companies have worked hard to keep the ratio of reserves to production (R/P) for crude oil at about 40 years, and for natural gas at about 70 years. However, in recent years few major new production fields have been found, and the exploration effort and cost required to maintain these ratios has been significantly increased. Ultimately, of course, supplies of oil and natural gas will be depleted to such an extent, or the cost of production will become so high, that alternative energy sources will need to be developed. In some regions of the world new production from non-traditional petroleum supplies, such as heavy oil deposits and oil-sands, are being developed to produce "synthetic" oil, and will be able to extend the supply of traditional crude oil. Coal is available in much greater quantities than either crude oil or natural gas, and the reserves to production ratio is much higher, currently on the order of 200 years. This ratio is sufficiently large to preclude widespread exploration for new coal reserves, although they are no doubt available. The challenges, however, of using coal in an environmentally acceptable manner, and for applications other than large-scale generation of electricity, are such that coal remains under-utilized.

Increasing concern about the long-term availability of crude oil and natural gas, and about the emission of greenhouse gases and pollutants from fossil-fuels, has led to increased interest in the use of coal to produce both gaseous and liquid fuels. Historically, coal was used to manufacture "producer gas" before the widespread availability of natural gas, and processes have also been developed to convert coal into synthetic forms of gasoline and diesel fuel. At the present time the commercial production of liquid fuels from coal is limited to South Africa, but other coal-producing countries are also now examining this as a possible option to replace liquid fuels derived from crude oil. Of course the greater utilization of coal in this way, or for the production of synthetic natural gas, would result in increased emission of greenhouse gases and other pollutants. As a result, there is also increasing research and development being conducted on so-called "carbon capture and storage," or "carbon sequestration" techniques. There are several proposed methods for separating the CO_2 which is released when coal is burned, or converted into synthetic liquid or gaseous fuels, and to store, or "sequester," this in some way so that it doesn't enter the atmosphere as a greenhouse gas. Proposals to date are at an early stage, particularly for the difficult CO_2 separation step, but there have been several pilot studies to establish the long-term storage of CO_2 in depleted oil and gas reservoirs. Other studies of the feasibility of storing large quantities of CO_2 in the deep ocean are also under way, but these are at a much earlier stage of development. If such carbon capture and long-term storage processes can be proven to be technically feasible and cost-effective, they could provide a way to expand the use of the very large coal reserves around the world, without undue concern about production of greenhouse gases.

At the present time our primary energy sources are dominated by non-renewable fossil fuels, with nearly 80% of global energy demand supplied from crude oil, natural gas, and coal. A more sustainable pattern of energy supply and end-use for the future will inevitably lead to the need for greater utilization of renewable energy sources, such as solar, wind, and biomass energy as well as geothermal and nuclear energy which many people consider to be sustainable, at least for the foreseeable future. Many assessments have shown that there is certainly enough primary energy available from renewable sources to supply all of our energy needs. Most renewable energy sources, however, have a much lower "energy density" than we are used to, which means that large land areas, or large pieces of equipment, and sometimes both, are required to replace fossil fuel use to any significant extent. This, in turn, means that the energy produced at end-use from renewable sources tends to be more expensive than energy from fossil fuels, even though the primary energy is "free." This is beginning to change in some cases, however, as fossil fuel prices continue to increase, and the cost of some renewable energy supplies, such as wind-power, drops due to improved technology and economies of scale. Other concerns with renewable energy arise due to their intermittent nature, however, and with the impact of large-scale installations, particularly in areas of outstanding natural beauty, or where there are ecological concerns.

Some observers are proposing the widespread expansion of nuclear power as one way to ensure that we have sufficient sources of clean, low-carbon, electricity for many generations to come. Although nuclear power currently accounts for nearly 7% of global primary energy supplies, there has been little enthusiasm for expansion of nuclear capacity in recent years. The lack of public enthusiasm for nuclear power appears to be primarily the result of higher costs of nuclear electricity production than was originally foreseen, as well as concerns over nuclear safety, waste disposal, and the possibility of nuclear arms proliferation. The nuclear industry has demonstrated, however, that nuclear plants can be operated with a high degree of safety and reliability, and has been developing new modular types of reactor designs which should be much more cost-effective than original designs, many of which date from the 1950s and 1960s. New nuclear plants are being built in countries with very high energy demand growth rates, like China and India, and electric utilities in the developed world are also starting to re-think their position on building new nuclear facilities. There will no doubt be a vigorous debate in many countries before widespread expansion of nuclear power is adopted, but it is one of the few sources of large-scale zero-carbon electricity that can be used to substantially reduce the production of greenhouse gases. The need for such facilities may increase if applications which have traditionally used fossil fuels, such as transportation, begin a switch to electricity as the energy carrier of choice, necessitating a major expansion of electricity generation capacity.

Transportation accounts for just over one-quarter of global energy demand, and is one of the most challenging energy use sectors from the point of view of reducing its dependence on fossil fuels, and reducing the emission of greenhouse gases and other pollutants. This is because the fuel of choice for transport applications is overwhelmingly gasoline or diesel fuel, due to the ease with which it can be stored on board vehicles, and the ubiquitous nature of the internal combustion engine which has been highly developed for over 100 years for this application. Although proposals have been made to capture and store CO_2 released during the combustion of fossil fuels in stationary applications, this is not a viable solution for moving vehicles of any kind. Hydrogen has been proposed as an ideal replacement for fossil fuels in the transportation sector, either as a fuel for the internal combustion engines now universally used, or to generate electricity from fuel cells on-board the vehicle. The use of hydrogen in either of these ways would result in near-zero emissions from the vehicle, of either greenhouse gases or other pollutants, and has been cited as an important step in developing the "hydrogen economy." If one looks at the complete energy conversion chain, however, it is clear that hydrogen is only the energy carrier in this case, and the primary energy source will necessarily come from either fossil fuels, or from renewable or nuclear sources, using electricity as an intermediate energy carrier. The use of renewable or nuclear energy as a primary source would result in zero emissions for the complete energy cycle, but the overall energy conversion efficiency would be very low, requiring a large expansion of the electricity-generating network. An alternative solution, with a much higher overall energy efficiency and lower cost, may be the successful development of "grid-connected," or "plug-in" hybrid electric vehicles, which use batteries charged from the grid to provide all of the motive power for short journeys, and a small engine to recharge the batteries if a longer range was required. In a later chapter we will examine these alternative transportation energy scenarios using the energy conversion chain approach.

The "energy problem," that is, the provision of a sustainable and non-polluting energy supply to meet all of our domestic, commercial, and industrial energy needs, is a complex and long-term challenge for society. Fortunately, man is by nature a problem-solving species, and there are many possible solutions in which future energy supplies can be made sustainable for future generations. The search for these solutions is, however, by its very nature a "multidisciplinary" activity, and involves many aspects of science, engineering, economics, and social science. The development of these solutions also tends to be very long-term, on the order of 10, 20, or even 50 years, and therefore far beyond the time-frame in which most politicians and decision-makers think. We must, therefore, develop new long-term methods of strategic thinking and planning, and make sure that some of the best minds, with a wide range of skills and abilities, are given the tools to do the job. This book summarizes the current state of the art in balancing energy demand and supply, and tries to provide some insight into just a few of the many possible scenarios to build a truly sustainable, long-term, energy future. No one individual can provide a "recipe" for energy sustainability, but by working together across a wide range of disciplines, we can make real progress towards providing a safe, clean, and secure energy supply for many generations to come.

2

The energy conversion chain

Every time we use energy, whether it's to heat our home, or fuel our car, we are converting one form of energy into another form, or into useful work. In the case of home heating, we are taking the chemical energy available in natural gas, or fuel oil, and converting that into thermal energy, or "heat," by burning it in a furnace. Or, when we drive our car, we are using the engine to convert the chemical energy in the gasoline into mechanical work to power the wheels. These are just two examples of the "Energy Conversion Chain" which is always at work when we use energy in our homes, offices, and factories, or on the road. In each case we can visualize the complete energy conversion chain which tracks a source of "primary energy" and its conversion into the final end-use form, such as space heating or mechanical work. Whenever we use energy we should be aware of the fact that there is a complete conversion chain at work, and not just focus on the final end-use. Unfortunately, many proposals to change the ways in which we supply and use energy take only a partial view of the energy conversion chain, and do not consider the effects, or the costs, that the proposed changes would have on the complete energy supply system. In this chapter we will discuss the energy conversion process in more detail, and show that some proposed "new sources" of energy are not sources at all, and that all energy must come from only a very few "primary" sources of energy.

A schematic of the global "energy conversion chain" is shown in Figure 2.1. Taking a big-picture view, this chain starts with just three "primary" energy sources, and ends with only a few end-use applications such as commercial and residential building heating, transportation, and industrial processes. Taking this view, our need for energy, which can always be placed broadly into one of the four end-use sectors shown on the far right in Figure 2.1, anchors the "downstream" end of

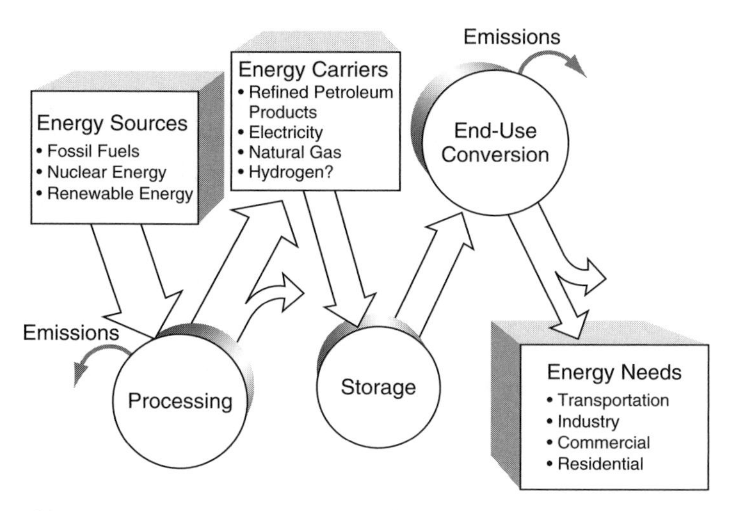

Figure 2.1 The energy conversion chain.

the conversion chain. This energy need is always supplied, ultimately, from one of the primary sources of energy listed on the far left-hand side of the diagram. In between the primary source and the ultimate end-use are a number of steps in which the primary source is converted into other forms of energy, or is stored for use at a later time. To take a familiar example, in order to drive our car, we make use of a fossil fuel, crude oil, as the primary energy source. Before this source provides the motive power we need, however, the crude oil is first "processed" by being converted into gasoline in an oil refinery, shown in the second step in Figure 2.1. The result of this processing step is the production of a secondary form of energy, or what is usually called an energy "carrier." Also, in this step there is usually some loss of energy availability in the processing step, as indicated by the branched arrow joining the processing block to the energy carrier block. There are, again, relatively few energy carriers, as shown in the third step of the diagram. Broadly speaking, these are refined petroleum products (gasoline in our car example), electricity, natural gas, and potentially, hydrogen. Once the primary source has been converted into the carrier of choice, it is usually stored, ready for later use in the final energy conversion step. In our automobile case, the gasoline is stored in the fuel tank of the vehicle, ready for use by the engine. When we start the engine, and drive away, the final step in the energy conversion chain is undertaken. This is the final end-use conversion step in which the chemical energy stored in the gasoline is converted into mechanical work by the engine to drive the wheels. In this step there are usually large losses of energy availability, due to the inherent inefficiencies of the end-use conversion step, and this is again indicated by the branched arrow in this step. If this step is representative of an automobile engine, for example, these energy losses may be on the order of two-thirds of the energy in the gasoline. This is, of course, just one example, but any energy-use scenario can always be followed through the complete energy conversion chain illustrated in Figure 2.1. In some cases, not all steps in the chain are required, but energy end-use can always be traced back to a primary energy source. For example, in most cases when electricity is the energy carrier it is used immediately upon production, due at least in part to the difficulty of storing electricity.

One striking lesson to be learned from Figure 2.1 is that there are only three primary sources of energy: fossil fuels, nuclear energy, and renewable energy. This means that every time we make use of an energy-consuming device, whether it is a motor vehicle, a home furnace, or a cell-phone charger, the energy conversion chain can be traced all the way back to one (or more) of these three main sources of primary energy. Also, in today's world there is currently very little use made of renewable energy (with the notable exception of hydroelectric power) as a primary energy source, so realistically we can almost always trace our energy use back to either fossil energy or nuclear power. And, finally, since nuclear power provides only a small fraction of the total electrical energy being produced today, fossil fuels are by far the most important source of primary energy. Fossil fuels can be broken down into three main sub-categories: coal, petroleum (or crude oil), and natural gas. Today, coal is a significant primary source of energy for electrical power generation, as is natural gas, while petroleum provides the bulk of the primary energy used to power our transportation systems. It can also be seen from Figure 2.1 that there are only three energy carriers that are of significance today; refined petroleum products, natural gas, and electricity. Hydrogen, often billed erroneously as an energy source of the future, is in fact an energy carrier, and not a primary source of energy. We shall discuss this issue in more detail in a subsequent chapter, but for the moment we simply show it as a possible energy carrier, as it is not presently used in this way to any significant degree.

Another important feature illustrated in Figure 2.1 is the release of emissions, both in the initial processing step and in the final end-use conversion step. Again using the automobile example, these are primarily in the form of carbon dioxide (CO_2), carbon monoxide (CO), unburned hydrocarbon gases (HCs), and nitrogen oxides (mainly NO and NO_2, but usually just described as NO_x). Some of these are released in the refining process, but most of them are released during the final conversion from chemical energy to useful work in the vehicle engine. This emission of pollutants from both the primary energy processing step, and the end-use step, provides an extremely important link between energy use and the environment. The reaction of unburned hydrocarbons and NO_x, in the presence of sunlight, for example, is responsible for smog formation, which has become a major problem in urban centers. This has been alleviated somewhat in the developed world by the introduction of stringent regulations to limit emissions from vehicles and power stations, but will continue to be a very serious problem with the growth in vehicle ownership, particularly in large developing economies.

The emission of CO_2, on the other hand, results in a quite different environmental problem; global warming brought about by the "greenhouse effect." We will discuss this effect in more detail in the next chapter, but will simply note here that the CO_2 molecules (and other greenhouse gases, such as methane) act like a selective screen, or "blanket," which allows short-wavelength radiation from the sun to pass through to warm the earth, but trap the longer wavelength energy which is normally re-radiated back out into space by the earth. This provides a net gain of energy by the earth's atmosphere, so that over time the global temperature increases. Although this has been somewhat controversial in the past, most scientists and observers now agree that global temperatures have increased by approximately 0.75 °C over the past 200 years, primarily due to anthropogenic, or man-made, increases in CO_2 concentration in the atmosphere. This concentration is some 370 parts per million (ppm) today, and has risen from a long-term average of 280 ppm before the industrial revolution of the eighteenth century. The Intergovernmental Panel on Climate Change (IPCC) has suggested that by the end of the twenty-first century the global concentration of CO_2 will be somewhere between 550 and 900 ppm, resulting in an increase in the average global temperature of between 1.4 and 5.8 °C. The consequences of such a large increase in average global temperature are somewhat uncertain, but it is quite likely that it would result in a shrinkage of the polar ice caps and a spread of severe drought conditions in some areas of the world. The IPCC has also suggested that the global mean sea-level could increase by between 0.1 m and 0.9 m by the end of the century, which, at least at the high end of the estimate, could have very serious consequences for coastal communities. Of course global warming could also mean an extension of the growing season in some parts of the world, so there may even be some positive benefits. The consensus appears to be, however, that any significant global warming would result in serious environmental degradation in many vulnerable parts of the world.

The energy storage block depicted in Figure 2.1 is not an energy conversion process, but it is a critical part of many energy systems. In many cases it is necessary to store the energy in its intermediate form as an energy "carrier" before the final end-use step. In such cases it is simply not practical to use the energy directly as it is produced in the initial conversion from primary energy to energy carrier. This is the case for the automobile, of course, as it would be completely impractical to feed a continuous supply of gasoline from the refinery to the vehicle's engine. The intermediate energy carrier is therefore stored after manufacture, often in several different stages, before ending up in the automobile's fuel tank. For example, gasoline is usually first stored in large tanks at the refinery, then transferred by delivery tanker trucks for secondary storage at filling stations, and finally pumped into the vehicle fuel tank when required. In fact, one of the major benefits of gasoline (or any liquid hydrocarbon fuel) is that it is easily stored, and has a very high "energy density," as we shall see later. Electricity, however, is quite difficult to store in large quantities, and it normally moves directly as an energy carrier to the final end-use conversion step. In this case the final end-use conversion is usually done by an electric motor, or a resistor-type heating element, and these are directly connected, through the electricity distribution system, to a generator at a power station. Because electricity can be moved through wiring efficiently over long distances, storage is not a requirement for fixed applications in our homes, offices, and factories. For transportation applications, however, other than for electric trains, or trolley buses, the storage of electricity is a major challenge. Batteries are very effective for small-scale application of electricity to devices such as laptop computers and other portable electronic devices, but do not yet have sufficient energy storage density for widespread application to electric cars, for example. We will examine this challenge in more detail in a subsequent chapter.

Another feature of the energy conversion chain is the loss of some "usable" energy during every processing step. Although the laws of thermodynamics tell us that energy is always conserved, and is neither created nor destroyed, some of it becomes unavailable to us at each step in the conversion chain. This "unavailable" (or "lost") energy usually ends up as low-temperature "waste-heat," and although this is still a form of energy, it is not technically or economically feasible to use it. If we again look at the case of the automobile, for example, usable energy is lost during the processing of crude oil in the refinery to produce gasoline, and again in the conversion of the chemical energy in the gasoline into useful mechanical work by the engine. This loss of usable energy, a consequence of the laws of thermodynamics, is usually quantified by an "efficiency" value, which is the ratio of usable energy produced, or work done, in an energy conversion process to the total energy available at the beginning of the process. In the case of the automobile the efficiency of conversion of crude oil into gasoline at the refinery is approximately 85%, while for conversion of the chemical energy in the fuel into mechanical work by the engine and drivetrain it is only about 20%. In other words, starting with 100 units of primary energy (usually measured in kilojoules, kJ) in the form of crude oil, we end up with 85 kJ of energy in the gasoline. When the gasoline is burned in the engine to produce mechanical power (the rate of doing work, measured in kW), this 85 kJ produces only 17 kJ (20% of 85 kJ) of useful work at the wheels. The overall energy efficiency of this process, from primary source to end-use, is therefore only 17%. The end result is that when we drive a typical car, some 83% of the primary energy ends up as "unavailable" energy, mostly in the form of low-temperature heat being rejected from the car radiator and exhaust gases, and from the refining process at the oil refinery.

This overall efficiency that we have just described, starting with the energy available at the primary source, and ending with the useful energy that we need to propel our car, or heat our homes and factories, is sometimes called the "well-to-wheels" efficiency, with obvious reference to the motor vehicle example we have just discussed. When comparing the performance of different approaches to meeting a particular end-use, whether it is an automobile, or a coal-fired powerplant, it is this "well-to-wheels" efficiency that is the best measure of the overall energy system performance. This efficiency describes the overall performance of the complete energy conversion chain, starting from the primary energy source and ending with the end-use application. A graphical illustration of this approach, using an "energy flow diagram," is sometimes very helpful, particularly for analyzing complex systems with multiple energy inputs and multiple end-uses. An example of such a diagram for the very simple case of the automobile that we have just discussed, is shown in Figure 2.2. The energy flow diagram, or Sankey diagram as it is often called, was first used by the

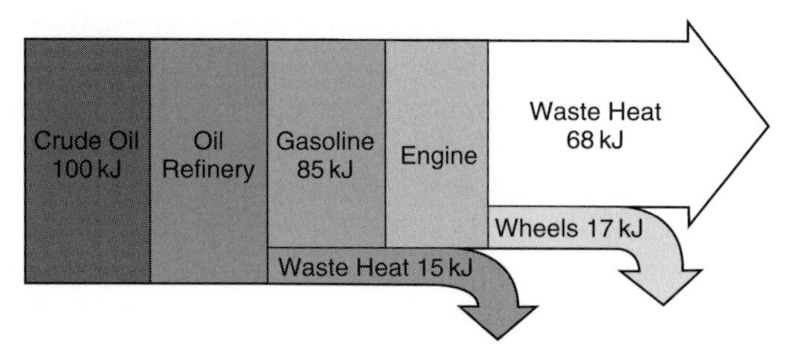

Figure 2.2 Simple energy flow "Sankey" diagram for an automobile.

nineteenth century Irish engineer, M. H. P. R. Sankey, to provide a quick visual representation of the magnitude of energy flows in the energy conversion chain. The two energy conversion steps for the case of an automobile using crude oil as a primary energy source are shown as boxes for the oil refinery, which converts crude oil into the gasoline energy carrier, and the engine which converts the chemical energy in the gasoline into mechanical work to drive the wheels. The width of the boxes or arrows representing energy flows are often drawn so that they are proportional to the fraction of total energy flowing in that direction.

A quick inspection of the diagram shows that for every 100 kJ of energy in crude oil that is used the refining process results in 85 kJ of available energy in the form of gasoline, and from this amount of energy the engine produces 17 kJ of useful work to drive the vehicle. The unavailable energy resulting from both these energy conversion steps is shown as "waste heat" in both cases. In the automobile, most of this waste heat is rejected to the ambient air from the hot exhaust gases and from the engine cooling water by the radiator. We can see, using this diagram as an example, that every time we use energy, our "end-use" is just one part of an extensive "energy conversion chain" leading back to one of only three primary energy sources. In order to understand the complete effects of our energy end-use on the environment, and on the long-term sustainability of the planet, we need to always consider the complete energy conversion chain. It is not good enough to simply analyze the "link" in the chain closest to our end-use if we are to fully understand the consequences of our energy choices. In subsequent chapters we shall begin to lay the groundwork to enable us to conduct a full "energy conversion chain analysis." We will also see the benefit of quickly being able to visualize energy flows using Sankey diagrams such as that shown in Figure 2.2 when we examine the complex flows from primary sources to end-uses for a complete energy economy. The Sankey diagram provides a very useful "snapshot" of the energy conversion chain, and clearly shows where energy is being lost, or converted into unavailable energy. Similar diagrams can be constructed to account for the total flow of energy, from primary sources to end-uses, for complete economies, or even for the total global energy consumption. These are particularly useful in showing the degree to which primary energy becomes "unavailable," or is lost in the form of waste heat. We shall discuss this more general form of the energy flow diagram in Chapter 10, when we look at global energy balances in more detail.

3

Energy and the environment

There is little doubt that the large-scale utilization of fossil fuels is putting significant stress on the environment. The effects of combustion products on air quality and the climate are both local and global in nature. The local effects, primarily in the form of air pollution and smog formation in large urban areas, have been known for many decades, and in recent years government regulations to reduce the effects of air pollution have been significantly strengthened. These include both exhaust emission standards for vehicles as well as emissions regulations for large fixed installations, such as fossil-fueled power stations. These regulations have been pioneered in the USA by agencies such as the California Air Resources Board (CARB), and the US Environmental Protection Agency (EPA), but similar measures have now been adopted in most of the developed world. On a global scale, there is increasing evidence, and concern, about the role of CO_2 and other so-called "greenhouse gases" on global climate change. In this chapter we will examine both the localized and global effects of these air emissions, and describe current mitigation techniques.

3.1 LOCALIZED ENVIRONMENTAL CONCERNS

Localized air pollution, prevalent in the heavily populated areas of large cities, results from direct chemical reaction with the products of combustion and from the formation of ground-level ozone. Combustion products include carbon monoxide (CO), sulfur dioxide (SO_2), nitrogen oxides (NO_x), unburned hydrocarbons, and finally carbon dioxide (CO_2), which is primarily of global concern. Carbon monoxide is a toxic gas which is usually formed in small concentrations from well-adjusted burners or internal combustion engines, but can be formed at higher levels if there is insufficient air present for complete combustion. In urban areas this is mainly a product of vehicle engine exhaust, although it has been greatly reduced by the widespread use of catalytic converters in car exhaust systems. As such, it is today rarely a threat to human health on its own. Sulfur dioxide is formed in the combustion process when fuels containing sulfur are burned, and this is now limited primarily to high-sulfur coal or in some cases to low-quality gasoline and diesel fuel containing high levels of sulfur. When SO_2 is released to the atmosphere from power station chimneys or vehicle exhausts it can react with water vapor to form sulfuric acid, an important component of "acid rain." In sufficient concentrations this can be very damaging to human lung tissue, as well as to buildings, vegetation, and the environment in general. The emission of SO_2 from coal-fired power stations, and subsequent acid rain formation, has been greatly reduced in recent years, however, by burning low-sulfur coal and by the installation of flue gas desulfurization (FGD) equipment. Emissions from vehicle exhausts have also been reduced by the on-going installation of sulfur removal equipment in oil refineries in order to remove sulfur from both gasoline and diesel fuel during the refining process.

Nitrogen oxides, NO and NO_2, collectively described as "NO_x," together with unburned hydrocarbons, are primarily a concern because of the potential to form ground-level ozone (O_3). Nitric oxide (NO) is formed during the combustion of fossil fuels in the presence of nitrogen in the air, whether in motor vehicles, thermal power stations, or in furnaces and boilers used to heat homes and commercial buildings. The NO formed during the combustion process is normally converted rapidly to NO_2 due to the presence of excess oxygen when it is discharged into the atmosphere. In the presence of sunlight, however, the NO_2 may subsequently be dissociated, resulting in the free oxygen atoms reacting with O_2 molecules to form high levels of "ground-level" ozone. Ozone is a very reactive oxidant and can cause irritation to the eyes and lungs, and can also destroy vegetation as well as man-made materials such as synthetic rubber and plastic. In high concentrations, found mainly in large urban centers with high levels of solar insolation and unburned hydrocarbons, it becomes "smog" with its characteristic brown color and odor. Smog contains a high concentration of highly reactive hydrocarbon free radicals, and not only causes visibility problems, but can result in severe health problems, particularly for people with asthma or other lung ailments. In response to environmental legislation in many parts of the world, techniques have been developed to significantly reduce the NO_x emissions from stationary combustion equipment such as boilers and large furnaces. The production of NO_x is directly related to the combustion temperature, and many companies have concentrated on reducing combustion temperatures, thereby reducing NO_x formation. This has resulted in the development of so-called "Low-NO_x" burners, which incorporate multi-staged combustion, or lean-burn technology in which excess air is used to reduce combustion temperatures. Where regulations are particularly stringent, a greater reduction in NO_x emission levels can be achieved by selective catalytic reduction, in which the reducing agent ammonia reacts with NO to produce nitrogen and water. For motor vehicles, the development of the three-way catalytic converter, which has the ability to both oxidize unburned hydrocarbons and CO, and reduce NO_x emissions, has been particularly effective in making modern vehicles much less polluting than has previously been possible. The introduction of the catalytic converter on gasoline vehicles has reduced the emission of NO_x by over 90% compared with a vehicle without the device.

In addition to the chemical effects of ozone and smog formation, there is increasing interest in the health effects of particulate emissions, which are primarily a feature of coal combustion and diesel engine exhaust. The particles are formed through a complex process involving unburned hydrocarbons, sulfur dioxide, and NO_x, primarily in fuel-rich flames such as those inherent in diesel engines and the pulverized coal combustion systems used in power stations. The particles formed have a wide size range, but the ones that have come under the most scrutiny for health reasons, and have been the subject of environmental legislation to limit their production, are those under 10 microns (1 micron $= 10^{-3}$ mm) in diameter. This so-called PM_{10} matter can enter deep into the lungs and there is growing scientific consensus that these can then cause serious heart and lung complaints, including asthma, bronchitis, and even lung cancer and premature death. Recently there has been increasing concern about the very smallest particles, $PM_{2.5}$, the material under 2.5 microns in characteristic diameter. There is some evidence that these may be of equal, or even greater, concern than the larger particles in that they have the ability to penetrate even deeper into the lungs. Particulate emissions from coal-fired power plants, which normally also include a significant fly ash content, have long been controlled by electrostatic precipitators, which use fine, electrically charged wires to attract the particulate matter, which is then periodically removed, usually by vibrating the wires. This technique tends to work well for large particle sizes, and in order to remove smaller size fractions the precipitator may be followed by a "bag-house," which is essentially a very large fabric filter. These techniques, however, are not sufficient for removing the very smallest particles, such as those produced by diesel engines. The removal of these very fine particles from diesel engines is particularly important in urban areas, where the population density is high, and people are in close proximity to diesel exhaust. In response to increasingly stringent regulations to limit the mass of particulate matter emitted by diesel engines, manufacturers have worked hard to reduce this by increasing fuel injection pressures. Ironically, some researchers have now expressed concern that this actually may have made matters worse, as the increased injection pressures result in much smaller particle sizes on average. The total mass of particulate matter emitted has been significantly reduced, but this has been achieved at the expense of producing many more of the very smallest particles. In recent years diesel engine manufacturers have been working to perfect a "particulate trap," to filter out the very fine particles contained in the exhaust gases. This is usually a very fine, porous, ceramic matrix which traps the particles but allows the gaseous exhaust products to pass through. After some hours of running the trap needs to be "regenerated," by burning off the entrapped particulate material. These devices have not yet been developed to the point where they are reliable enough, or inexpensive enough, to be routinely fitted to commercial vehicles.

3.2 GLOBAL ENVIRONMENTAL CONCERNS

On a global scale, it is the "greenhouse effect" and the prospect of global warming which has drawn the most attention. A simple diagram illustrating this effect is shown in Figure 3.1. Solar radiation produced as a result of the very high temperature of the sun is composed primarily of short wavelength visible or near-visible radiation, for which the atmosphere is largely "transparent." In other words, although a small fraction of this radiation is reflected by the earth's atmosphere back out into space, most of it passes straight through (as if the atmosphere is window glass) and warms the earth's surface. The warm earth then re-radiates some of this energy back out into space, but since it is produced at relatively low temperatures it is primarily long wavelength, or infra-red radiation. Some of the gases in the earth's atmosphere, just like window glass, are particularly opaque (or have a low "transmissivity") to this long-wavelength radiation, and are therefore referred to as "greenhouse gases" (GHGs). Much of the long wavelength radiation is therefore reflected back to the earth's surface and

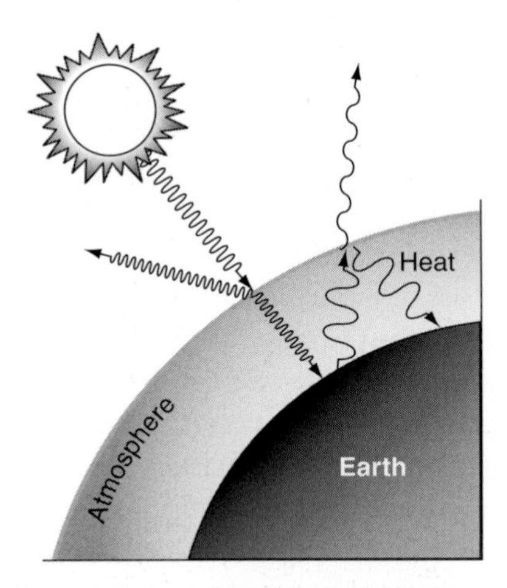

Figure 3.1 The atmospheric "greenhouse" effect.

there is then a net imbalance in the energy absorbed by the earth and that re-radiated back out, with the result being a warming of the earth's surface and the surrounding atmosphere, just as in a greenhouse.

The degree of this energy imbalance depends very much on the transmissivity of the atmosphere, in other words the degree to which the gases in the atmosphere either transmit or block the infra-red radiation from the earth. Climatologists refer to the effects of changes in the amount of solar radiation reaching the earth's surface as changes in the "radiative forcing" of the atmosphere. Some gases are much more opaque to the long wave-length radiation leaving the earth's surface than others, and their relative effect is measured by their "global warming potential" (GWP). Probably the most important of these gases is water vapor, and its concentration in the atmosphere can vary significantly, both spatially and temporally. However, the amount of water vapor in the atmosphere is primarily a function of natural processes, and it is therefore not usually considered to be an anthropogenic (man-made) GHG. The atmospheric gases which are anthropogenic in nature, and which have increased in concentration over time, include carbon dioxide (CO_2), methane (CH_4), nitrous oxide (N_2O), and a variety of gases, such as the chlorofluorocarbons (CFCs), which exist in small quantities, but have a strong global warming potential. Since CO_2 exists in the atmosphere in much greater quantity than the other anthropogenic GHGs, it is usually assigned a GWP rating of 1.0. The two next most important GHGs are CH_4, with a GWP of 23, and N_2O, with a GWP of 296 (see Houghton, 2004). Even though CO_2 has the lowest GWP of the three gases, it is by far the most important

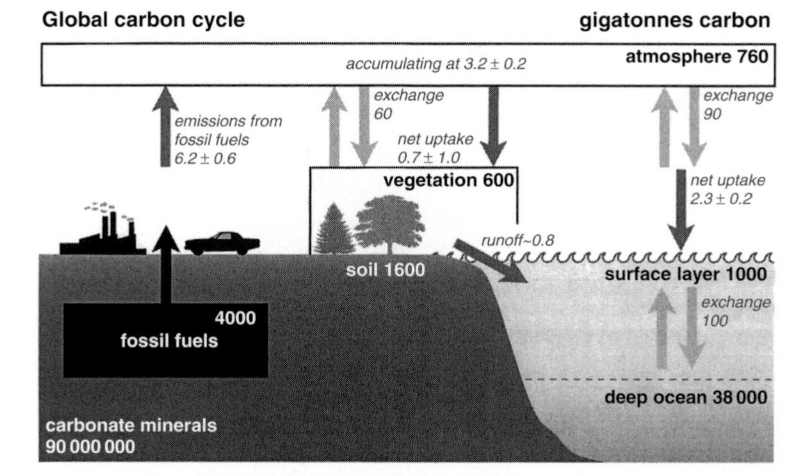

Figure 3.2 The global carbon cycle. *Source*: Royal Commission on Environmental Pollution's 22nd Report: *Energy – The Changing Climate*.

because it is emitted in much greater quantity. Houghton estimated that CO_2 has accounted for some 70% of the enhanced greenhouse effect resulting from the anthropogenic release of GHGs, while methane accounts for 24%, and N_2O for 6%. For this reason CO_2 has received the most attention from scientists and policymakers, although it is not the only GHG of importance. If over time the long-term average concentration of CO_2 in the atmosphere increases, there will be a decrease in the long wavelength transmissivity of the atmosphere, resulting in more of the infra-red radiation being trapped. This will lead to an increase in the net energy being absorbed by the earth's surface and the atmosphere, with the result being an increase in the global average temperature. There is, therefore, increasing scrutiny of the "global carbon cycle" and a concern with increasing concentration levels of CO_2 in the atmosphere.

The "global carbon cycle," illustrated in Figure 3.2, taken from the report of the UK Royal Commission on Environmental Pollution, *Energy – The Changing Climate* (2000), shows the quite complex processes at work exchanging carbon between different parts of the earth and its atmosphere. The bold figures in each "reservoir" represent the amount of carbon stored, in units of gigatonnes (Gt – or billions of tonnes). The gray arrows represent natural exchanges between reservoirs, which are nearly in balance, while the bold arrows represent the net flux in each case. The figures in italics adjacent to each of the arrows show the

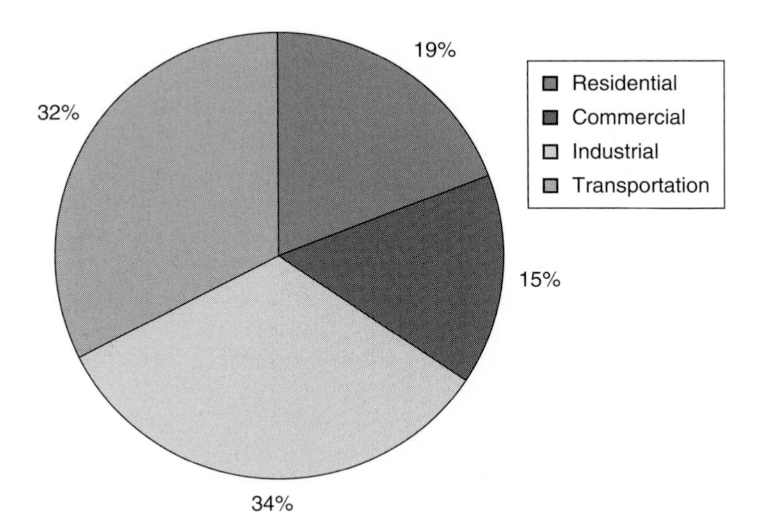

Figure 3.3 Emissions of CO_2 in the USA by sector, 1995. *Source*: Based on figures from the Energy Information Agency *Emissions of Greenhouse Gases in the United States 1995.*

CO_2 fluxes, in units of Gt/year of carbon, between the different reservoirs. It is clear that the natural fluxes are much greater than the anthropogenic flux resulting from the combustion of fossil fuels and industrial processes such as the production of cement. The net result of all of the net carbon fluxes shown is an accumulation of approximately 3.2 Gt/year of carbon in the atmosphere. In addition to carbon stored as CO_2, there is approximately 4000 Gt of carbon stored as fossil fuels; coal, oil, and natural gas, in the earth's crust, as shown in Figure 3.2. It is the consumption of these resources that is the main source of the anthropogenic release of some 6.2 Gt/year of CO_2 into the atmosphere. The fossil fuel reserves are relatively modest compared with the amount of carbon stored in the oceans, or in the earth as carbonate minerals, but are also much greater than the total carbon in the earth's atmosphere. They do, therefore, represent a substantial potential source of carbon which would be added to the atmosphere if they were all to be eventually consumed to provide mankind's energy needs without capturing and storing the CO_2 released.

The combustion of fossil fuels is the primary source of CO_2 emissions, and as such can be traced back to the major energy end-use sectors, including residential and commercial buildings, industrial processes, and transportation. Figure 3.3 shows the distribution of

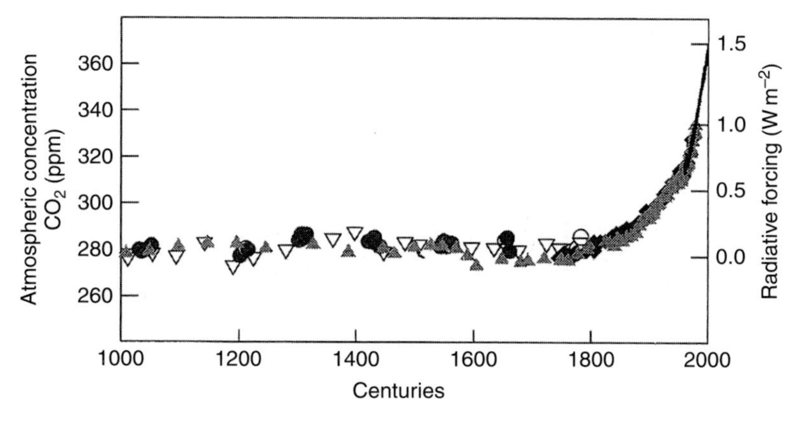

Figure 3.4 Atmospheric CO_2 concentrations. *Source*: IPCC *Climate Change 2001: The Scientific Basis.*

CO_2 emissions by end-use sector in the USA for the year 1995. Contributions from each end-use sector naturally vary from one country to another, depending on the state of industrial development, and particularly on the number of motor vehicles in operation. In the highly industrialized countries, for example, transportation, industrial processes, and electric power generation tend to be the dominant users of fossil fuels, and therefore also the dominant sources of CO_2 emissions. Nearly 35% of the total emissions shown in Figure 3.3, for example, originate from electrical powerplants. In less-developed nations, fossil fuel use, and therefore CO_2 emissions, may be heavily weighted towards domestic heating and cooking, rather than to the use of motor vehicles. In some sectors the use of fossil fuels, and therefore CO_2 emissions, can be reduced by switching from a high-carbon content fuel like coal, to a lower carbon content fuel, such as natural gas. This has been done in parts of Europe, for example, where coal-fired power stations have been replaced by natural gas-fueled combined cycle gas turbines (CCGTs). Also, increasing the end-use efficiency in any sector can be effective in reducing energy consumption, thereby reducing CO_2 emissions. This increase in efficiency may be easier to achieve in some sectors, for example domestic home heating, than in others, such as transportation. However, the introduction of fuel efficiency standards for motor vehicles in the USA, as well as increased fuel costs and switching from gasoline to more efficient diesel engines in some markets, has led to significant gains in the efficiency of automobiles over the past three decades.

Figure 3.4, from the Intergovernmental Panel on Climate Change, or IPCC (2005), shows the concentration of CO_2 in the

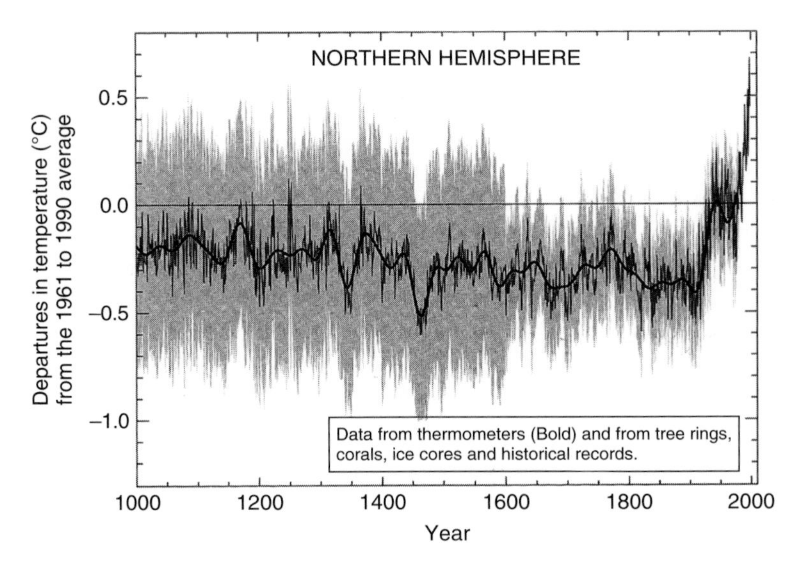

Figure 3.5 Earth's surface temperature change. *Source*: IPCC *Climate Change 2001: The Scientific Basis.*

atmosphere over the last 1000 years. It can be seen that the CO_2 concentration prior to the industrial revolution, beginning in the late eighteenth century, was nearly constant at a level of 280 parts per million (ppm). During the nineteenth and twentieth centuries the level has increased rapidly, reaching approximately 370 ppm today. This concentration represents the total carbon content of some 760 Gt currently in the atmosphere, as shown in Figure 3.2.

The effect of this large increase in CO_2 concentration on the earth's surface temperature can be seen in Figure 3.5, with data from various sources, including thermometer measurements over the past two centuries, and temperatures inferred from tree rings, ice cores, and other historical records for earlier times. It can be seen that there is a very good correlation between the increase in global CO_2 concentration (as seen in Figure 3.4) and the increase in the earth's temperature.

Scientists working with the Intergovernmental Panel on Climate Change (IPCC) have also done extensive computer modeling of the greenhouse gas effect to try to predict the effect of further increases in CO_2 concentration levels on global average temperatures. The computer models have used a number of different emissions and economic activity scenarios in order to better estimate the likely range of CO_2 concentration and average global temperature rise. The results of these calculations show that CO_2 concentration will likely reach a value

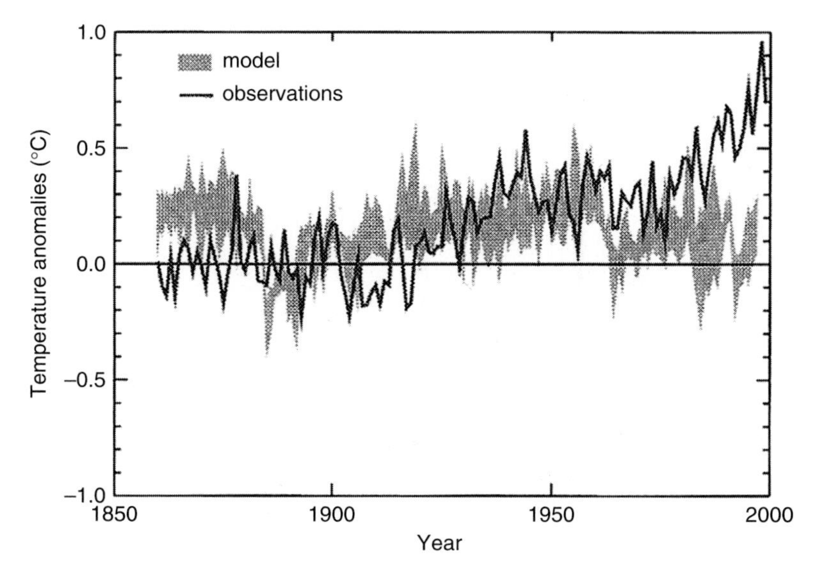

Figure 3.6 Predicted temperature change, natural forcing only.
Source: IPCC *Climate Change 2001: The Scientific Basis*.

ranging between about 550 and 900 ppm by the end of the twenty-first century, depending on the particular scenario chosen. These models have also examined the relative effect on global temperature of "natural forcing" of the atmosphere, due to variations in solar output, for example, and the so-called "anthropogenic forcing" due to man-made emissions of greenhouse gases. Figures 3.6 to 3.8 show the results of the model predictions for a base-case scenario compared with measured values of the temperature change from 1850 to 2000. The model predictions have been conducted first with the assumption of only natural forcing, then with only anthropogenic forcing, and finally with both natural and anthropogenic forcing, as shown in Figures 3.6 to 3.8 respectively.

In Figure 3.6, it can be seen that there is quite a poor correlation between the predicted temperature rise, assuming only natural forcing, and the rise obtained from actual observations. This is particularly true for about the first 25 years when the industrial revolution was well under way, and for the last 25 years during which there has been strong economic activity in many countries, with a consequent substantial increase in CO_2 emissions. With the assumption of only anthropogenic forcing in the model, as shown in Figure 3.7, the prediction is much better during the early and late years, but not very good

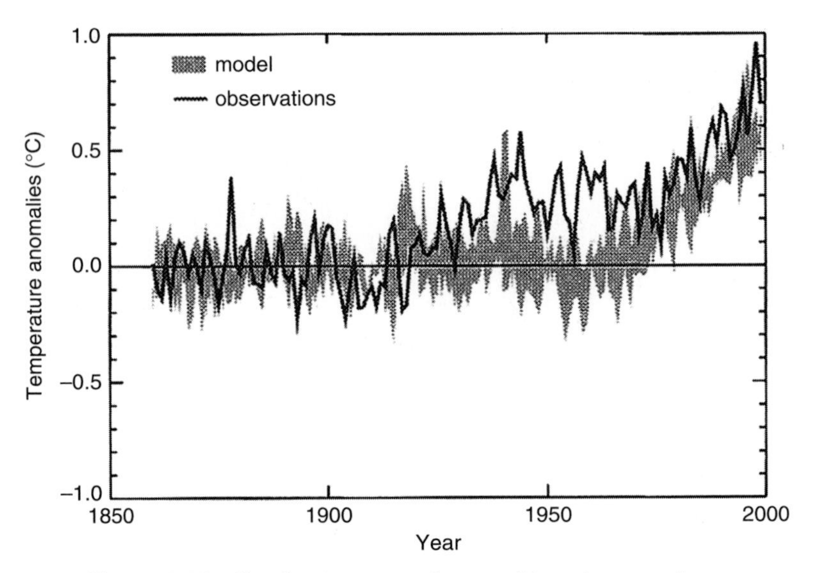

Figure 3.7 Predicted temperature change with anthropogenic forcing only. *Source*: IPCC *Climate Change 2001: The Scientific Basis.*

during the two decades between 1950 and 1970, when there was a noticeable decrease in solar intensity.

Finally, by including the effects of both natural and anthropogenic forcing in the model, the predicted temperature rise, as shown in Figure 3.8, matches very closely with the observed temperature records. The results from these three sets of predictions provides very strong evidence that the rapid increase in temperature observed over the last 50 years is very likely due to anthropogenic effects, and can be almost entirely attributed to the burning of fossil fuels.

Although the model results shown in Figures 3.6 to 3.8 were obtained for the base-case emissions and economic activity scenario, calculations were also conducted by researchers for a range of alternative scenarios, known as SRES (Special Report on Emissions Scenarios), as described by the IPCC. The results of predictions for the next 120 years for the complete range of these scenarios are shown in Figure 3.9. Results are shown both for a full range of several models using all of the SRES scenarios, and for a more restricted ensemble of models, again using the full range of SRES scenarios. These calculations predict an overall increase in global average temperature between 1990 and 2100 to range from a low of 1.4 °C to a high of 5.8 °C. At the higher end of this range, there would no doubt be significant

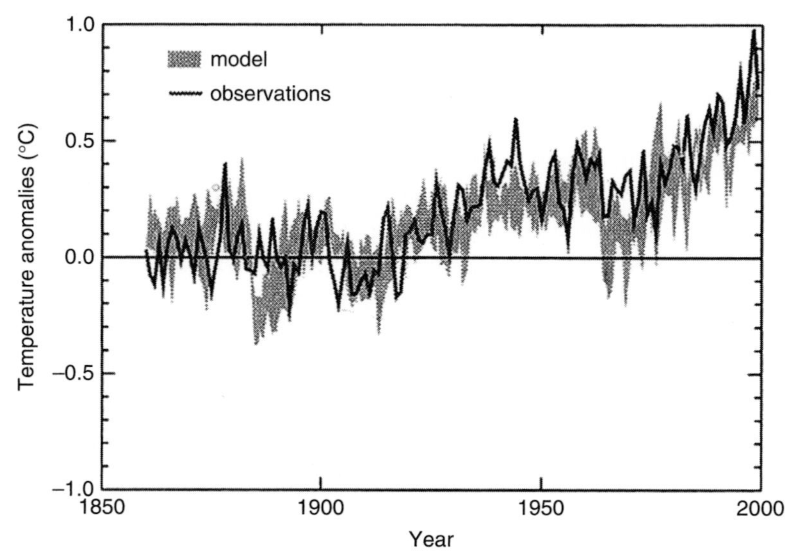

Figure 3.8 Predicted temperature change from both natural and anthropogenic forcing. *Source*: IPCC *Climate Change 2001: The Scientific Basis.*

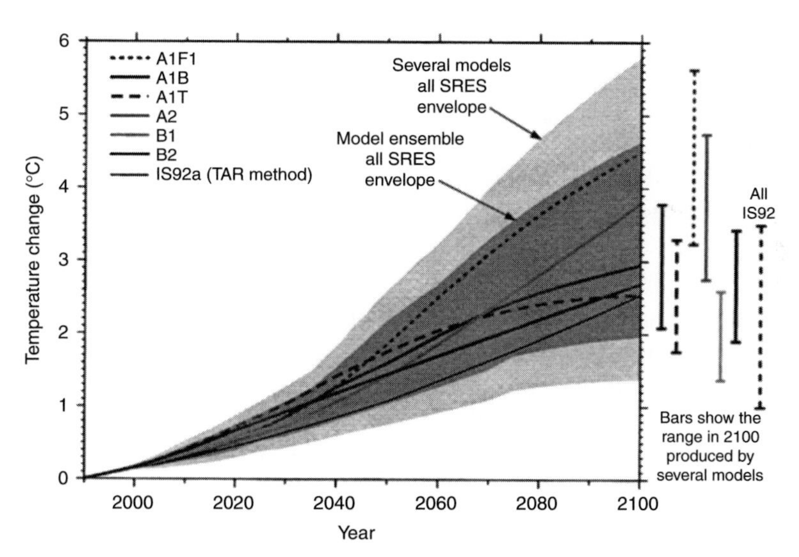

Figure 3.9 Temperature change predicted for various emission and economic activity scenarios. *Source*: IPCC *Climate Change 2001: The Scientific Basis.*

changes to the global climate, including more frequent and more severe storms, melting of the polar ice caps, and more frequent occurrence of droughts. There would also be a significant rise in mean sea level, predicted to be up to one meter, leading to widespread erosion and flooding in coastal areas worldwide.

Given the real threat that such climate change would have on mankind's well-being, and on the global economy, scientists, engineers, and policymakers are now discussing long-term mitigation techniques to minimize, or at least reduce, the rapid increase in global CO_2 concentration levels being predicted for the twenty-first century. At the present time these discussions are primarily focused on obtaining international agreement for limiting the production of greenhouse gases under the auspices of the United Nations Framework Convention on Climate Change (UNFCCC), which was formally adopted in 1992 in New York. Under this convention the most heavily industrialized countries, including the OECD members and 12 countries with "economies in transition," sought to return their greenhouse-gas emissions to 1990 levels by the year 2000. This was followed by the "Kyoto Protocol," committing signatories to specific action, which was proposed in Kyoto, Japan in 1997. Under this agreement, the industrialized countries listed in Annex 1 to the Kyoto accord agreed to reduce their emissions of a suite of six greenhouse gases below the levels produced in 1990 by "targets" of between 0% and 8%, averaged over the period from 2008 to 2012, as shown in Table 3.1. In one or two special cases

Table 3.1. *Kyoto accord targets*

"Annex 1" countries	Target %
European Union-15, Bulgaria, Czech Republic, Estonia, Latvia, Lithuania, Romania, Slovakia, Slovenia, Switzerland	−8
USA[a]	−7
Canada, Hungary, Japan, Poland	−6
Croatia	−5
New Zealand, Russian Federation, Ukraine	0
Norway	1
Australia[a]	8
Iceland	10

Note:
[a] The USA and Australia did not ratify the agreement.
Source: Houghton, 2004.

(Australia, Iceland, and Norway) the agreed targets were actually an increase from the 1990 levels due to the difficulties for smaller economies in making the necessary changes to their energy supply system. The 15 European Union (EU) countries (prior to the admission of 10 new countries in 2004) agreed to obtain the 8% reduction in GHG emissions on average across the whole community. Subsequent negotiations between the EU countries has resulted in the UK target for the period 2008–2012 being set at 12.5% below the 1990 level, for example.

The Kyoto Protocol was subject to final ratification by the countries who were "parties" to the convention, and the protocol was to enter into force on the 90th day after the date on which not less than 55 parties to the convention, incorporating Annex 1 countries which accounted in total for at least 55% of the total GHG emissions for 1990 from that group, submitted their final ratification notice to the UN. The USA and Australia subsequently went on record as saying that they would not ratify the agreement. As of November 2, 2004, 127 states and regional economic integration organizations had submitted their notice of ratification, and the total percentage of emissions from 'Annex 1' countries ratifying the agreement was 44.2%. The protocol would therefore come fully into force if either the USA, which accounts for 36.1% of Annex 1 emissions, or the Russian Federation, which accounts for 17.4% of these emissions, were to submit their notice of ratification. Although the USA indicated that they would not ratify, the President of Russia signed a federal law to ratify the protocol on November 4, 2004. The Kyoto Protocol then came into force on February 16, 2005, 90 days after Russia's notice of ratification was received by the UN in New York.

Whether or not most countries will actually meet these targets is in considerable doubt, particularly given the fact that very few mitigation techniques have been established to date. Also, the fact that the world's largest economy has not signed on to the agreement raises the issue of industrial competitiveness among those countries which do undertake mitigation measures. This will be particularly important given the growing economic power (and greenhouse gas emissions) of the rapidly growing economies, such as China and India, which have not been a party to the Kyoto Protocol. Although some countries (such as the UK) have made considerable progress in meeting their Kyoto targets, this has largely been the result of widespread "fuel switching" from coal to natural gas for electrical power generation. The fuel share for gas- and coal-fired power generation in the UK are now just about equal, at nearly 40% each, while in 1990 there was essentially no

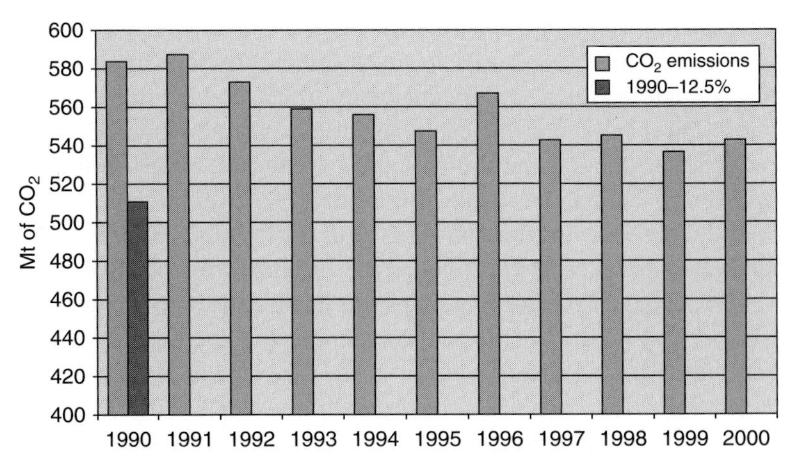

Figure 3.10 CO_2 emissions in the UK – 1990–2000. *Source*: UNFCCC.

gas-fired power generation. Figure 3.10 shows the annual CO_2 emissions for the UK, over the period 1990 to 2000, as reported to the UNFCCC. Given that CO_2 represents about 70% of the potential global warming effect of all GHG emissions worldwide, it is an important marker for achieving the Kyoto targets. The Kyoto target for the UK of 12.5% below 1990 levels for all GHGs is also shown on the left-hand side of Figure 3.10 in terms of a CO_2 emissions target. It can be seen that with the exception of one or two years, there has been a steady decrease in CO_2 emissions over the decade (colder than normal temperatures during the winter of 1995–96 resulted in the spike in emissions for 1996). The emission levels appear to have started to move up again in 2000, but it is too soon to know whether or not this is a long-term trend and whether the UK's Kyoto target will be achieved in the 2008–2012 time-frame of the accord.

The widespread fuel switching has been done primarily because of the lower capital cost of building natural gas-fired power stations compared with coal-fired ones, and also due to the relatively low cost of natural gas at the time these plants were constructed during the 1990s. The inherently lower carbon content of natural gas, coupled with the significantly increased efficiency of the combined cycle gas turbine power stations, has resulted in a large reduction in CO_2 emissions. With natural gas prices rising rapidly in recent years, however, and with concerns about shortages of natural gas supplies, this

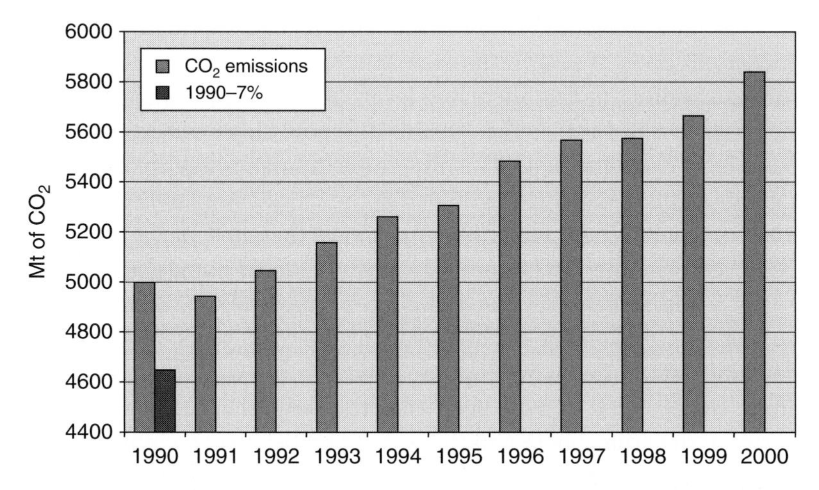

Figure 3.11 CO_2 emissions in the USA – 1990–2000. *Source*: UNFCCC.

fuel-switching may turn out to have been a poor short-term decision. The large-scale use of natural gas for electrical power generation may also come to be seen as very short-sighted in coming decades when natural gas, a premium fuel for heating commercial and residential buildings, for example, becomes harder to find and its price increases substantially as a result. In some countries, such as Canada, which relies on hydroelectric power for much of its electricity production, there is much more limited scope for fuel switching, and very little progress on meeting Kyoto targets has been made. In the USA, although there has been some fuel switching to natural gas, this has not been as widespread as in the UK, due in part to limited natural gas resources and increased prices. The annual CO_2 emissions in the USA over the period 1990–2000, as reported to the UNFCCC, are shown in Figure 3.11. The Kyoto 'target' was established in early negotiations at 7% below 1990 levels by 2008–2012, even though in the end the agreement was not ratified by the USA. It can be seen that there has been a steady increase in emissions over this 10-year period, so that by 2000 the CO_2 emissions were some 17% *above* the 1990 level. This means that the production of greenhouse gases in the USA would have to be reduced by some 20% from the 2000 level of 5840 million tonnes (Mt) to reach the 7% below 1990 target level of 4649 Mt, averaged over the years 2008–2012. Given the short period of time remaining, it seems very unlikely that the Kyoto target could be achieved without a serious impact on US global competitiveness.

Finally, we should note that not all scientists share the view that increased levels of CO_2 in the atmosphere are due primarily to mankind's activities, or that increased levels of CO_2 in the atmosphere are necessarily a bad thing. The 1995 IPCC report did conclude that "the balance of evidence suggests a discernible human influence on global climate," and this was strengthened in the Third Assessment Report of 2001 to "there is new and stronger evidence that most of the warming observed over the last 50 years is attributable to human activities." What is indisputable is that the concentration of CO_2 in the atmosphere has been rising steadily over the past 200 years, after having remained static for many centuries, as seen in Figure 3.4. However, from Figure 3.2 it can be seen that the natural exchanges of CO_2 between the atmosphere and the ocean, and between the atmosphere and global vegetation cover, are about a factor of 10 times the anthropogenic release of CO_2. One might conclude, therefore, that small perturbations in these natural exchanges could be just as important in determining the net CO_2 uptake of the atmosphere as the anthropogenic contributions. Some scientists have even claimed that because CO_2 is a natural "fertilizer" for vegetation, increased levels will enhance the global production of biomass, and make a positive contribution to the well-being of the planet. While these ideas do not represent the prevailing world view, there has been much discussion about the real cause of global warming, and it appears that more research is needed before we can definitely conclude that increasing global temperatures are the result of mankind's activities. Unfortunately, given that the world's largest economy, the USA, has stated that it will not ratify the Kyoto agreement, the debate often appears to be more political than scientific in nature.

3.3 ADAPTATION AND MITIGATION

Both adaptation, in which mankind simply learns to adapt to a changing global climate, and mitigation, in which measures are taken to limit CO_2 emissions, have been proposed. Although adaptation has not been a strategy favored by the majority of scientists, engineers, and policymakers, there is an argument that says that mankind is adaptable, and can always learn to live under new circumstances, if they are not too harsh and do not occur too rapidly. For example, while an increase in global average temperature of between 1.4 and 5.8 °C over the next 100 years has been predicted by the IPCC, some would argue that overall this change would not necessarily be catastrophic for mankind. While at the higher end of this range life could become intolerable for those living in or near desert regions, it could also result in an extension of the growing season in parts of Northern Canada and Russia, for example, leading to an increase in the ability to grow additional crops for both food and the production of biomass fuels. Although few definitive studies have been done, those advocating an adaptation strategy would say that the cost, in both human and financial terms, of adapting and perhaps moving people around the globe, could be less than the cost of significantly reducing global CO_2 emissions.

In the meantime, while the scientific debate over global warming continues, there are a number of mitigation, or "carbon abatement," measures that countries worldwide are taking to try to limit CO_2 emissions. Ratification of the Kyoto Protocol will act as the "carrot" to ensure that signatories to the agreement establish policies and procedures aimed at achieving their target reductions by the 2008–2012 time-frame. Some of the carbon abatement measures being taken, or at least being considered, include the introduction of stricter fuel consumption standards for new automobiles, measures to increase the efficiency of energy-intensive industrial processes and thermal power generation, and energy conservation measures such as improved insulation for houses and commercial buildings. Fuel switching, from coal-fired to natural gas-fired power generation, or even to nuclear power generation, can also produce large reductions in CO_2 emissions. In some cases reforestation is being intensified not only to replace trees lost to timber production, but to enhance the role of the global biomass as a CO_2 "sink." However, as we have seen, for most industrialized countries it will be difficult to meet their Kyoto targets in the relatively short time remaining before the end-dates of 2008–2012.

BIBLIOGRAPHY

California Air Resources Board (2005). *http://www.arb.ca.gov/homepage.htm*

Houghton, J. (2004). *Climate Change – The Complete Briefing*. Cambridge University Press.

Intergovernmental Panel on Climate Change (1995). *IPCC Second Assessment – Climate Change 1995*. Geneva, Switzerland: IPCC.

Intergovernmental Panel on Climate Change (2001). *IPCC Third Assessment Report: Climate Change 2001* (eds. Watson, R. T. and the Core Writing Team). Geneva, Switzerland: IPCC.

Intergovernmental Panel on Climate Change (2005). *http://www.ipcc.ch/*

International Energy Agency (2005). *http://library.iea.org/index.asp*

Kasting, J. (1998). The carbon cycle, climate, and the long-term effects of fossil fuel burning. In *Consequences: The Nature and Implication of Environmental Change*, **4** (Number 1).

Royal Commission on Environmental Pollution (2000). *Energy – The Changing Climate*. 22nd Report.

United Nations Framework Convention on Climate Change (2005). *http://www.unfccc.int/*

US Department of Energy. Energy Information Agency. *http://www.eia.doe.gov/*

US Environmental Protection Agency. *http://www.epa.gov/*

Part II The global energy demand and supply balance

4

World energy demand

We use energy in several different forms as we go about our daily lives, and rarely stop to think about the consequences of doing so. Energy is needed not only for our domestic needs, but also to fuel our factories and provide the motive power for transportation, whether that is by road, rail, air, or sea. The total world energy consumption in 2002 was just over 400×10^{15} Btu (British thermal units), or 10 Gtoe (Billion tonnes of oil equivalent). The distribution of that demand, by economic sector, is shown in Figure 4.1. Although the distribution by economic sector varies widely from country to country, depending largely on the degree of industrialization, overall approximately 25% of all energy is used to provide transportation, 32% is used to fuel industrial operations, while the balance is used for a range of activities, including the heating of both public and private buildings. A small quantity of primary energy resources are also used for so-called "non-energy" uses, such as chemical feedstocks used to produce plastics.

Although it is difficult to obtain a more detailed breakdown of energy consumption by economic sector on a global basis than that shown in Figure 4.1, most industrialized countries provide separate data for both the commercial and residential sectors. Energy demand by economic sector for the USA in 2000, for example, is shown in Figure 4.2. It can be seen that the share of energy used in the transportation and industrial sectors is just slightly greater than the world share. The demand for energy to supply heating and cooling, and to operate household appliances, in the residential sector is nearly equal to that needed to provide heating and cooling and operate office equipment in commercial buildings.

Worldwide demand for energy has been steadily increasing over time since the beginning of the industrial revolution in the eighteenth

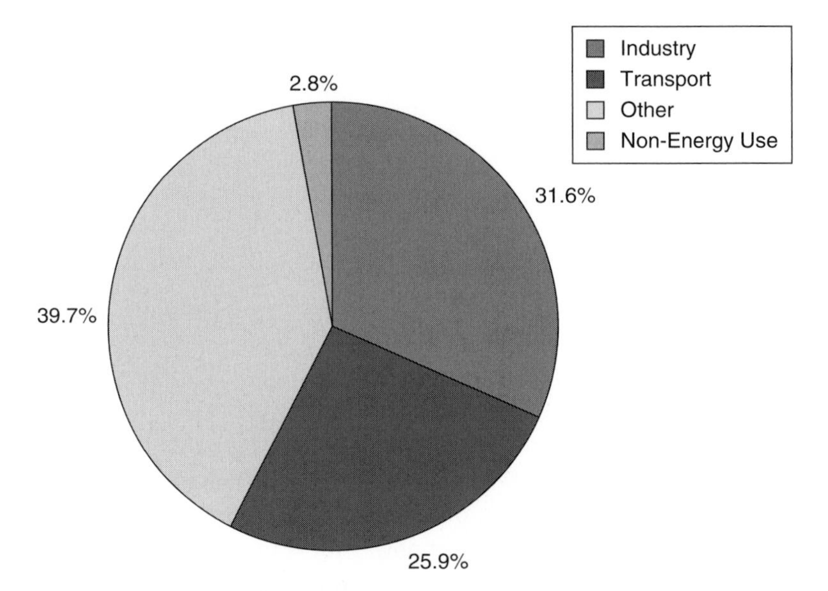

Figure 4.1 World energy demand by economic sector – 2002. Other – Includes commercial and residential buildings. *Source*: Based on figures from the International Energy Agency *World Energy Outlook*.

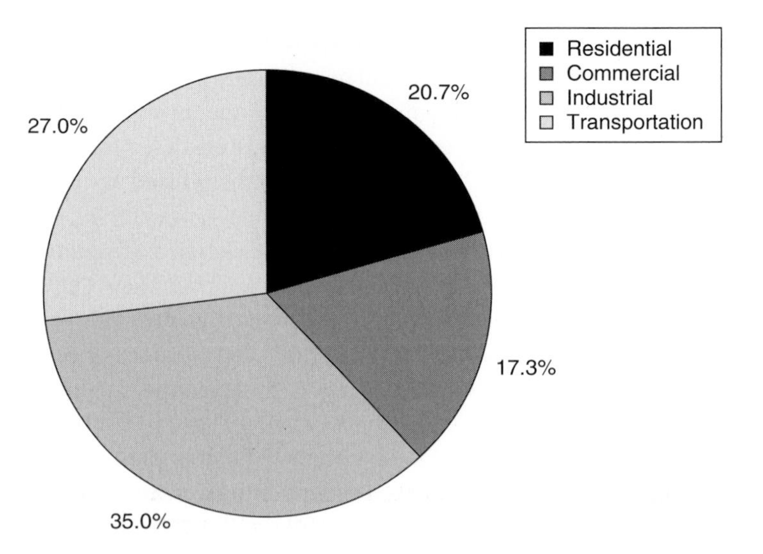

Figure 4.2 US energy demand by economic sector – 2000. *Source*: Based on figures from the Energy Information Administration *Annual Energy Review*.

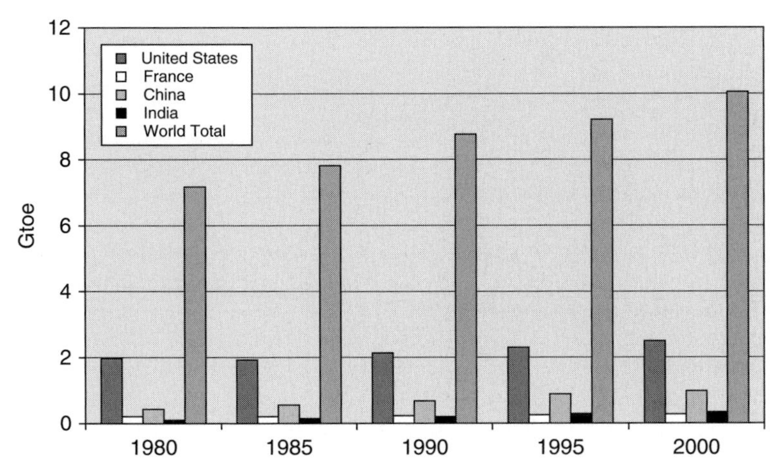

Figure 4.3 Growth in total energy demand – 1980–2000 (Gtoe).
Source: Based on figures from the Energy Information Administration *International Energy Annual 2002*.

century. The evolution of this increase in demand is shown for the 20 years between 1980 and 2000 in Figure 4.3 for the world as a whole, and for a few selected countries. The global consumption of all primary energy forms in 2000 was approximately 10 Gtoe, with the USA accounting for some 25% of the total worldwide demand. The growth in worldwide energy demand for this short period has been some 40%, as can be seen in Figure 4.3, although the growth rates in some individual countries have been much higher. While the total demand for energy in absolute terms from China and India, for example, has been much less than that of the USA, the growth rate in demand over this period has been very high.

The normalized growth in demand (with demand in 1980 set to 1.0) for the same countries over this 20-year period can be seen in Figure 4.4, compared with the world total growth in demand. It can be seen that the major industrialized countries, such as the USA and France, have had fairly modest growth during this period, but in the two largest "developing" countries, China and India, growth has been much higher. The average annual compound growth rate over this 20-year period has been 1.2% in the USA, and 1.75% for the world overall, but in China it has been 4.1% and in India 6.0%. These very high growth rates for the large emerging economies will put enormous pressure on worldwide energy resources in the years to come. If the 6% growth rate were to be sustained in India, for example, the total

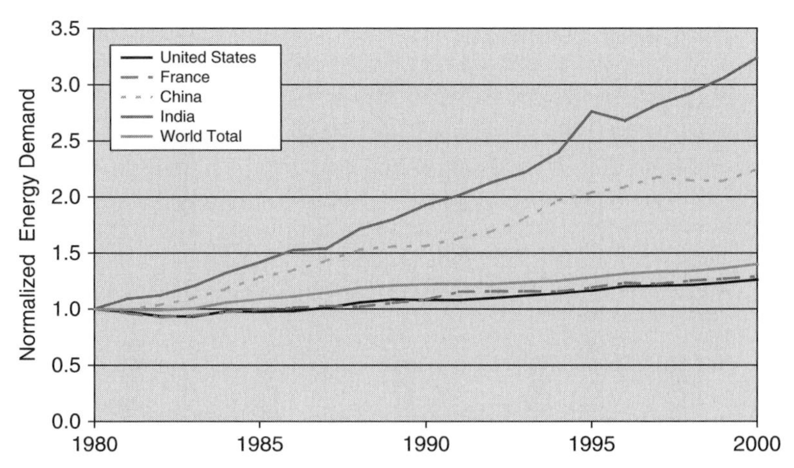

Figure 4.4 Normalized total energy demand – 1980–2000 (1980 = 1).
Source: Based on figures from the Energy Information Administration *International Energy Annual 2002*.

demand for energy would double every 12 years. If these rates were to continue until the middle of the century, China would overtake the USA as the largest energy consumer by the year 2030, and India would similarly overtake the USA by 2043. High energy demand growth rates will also put increasing strain on the global environment, unless action is taken to significantly change energy end-use patterns and the way we use our primary energy resources.

There has been, historically, a strong link between Gross Domestic Product (GDP) and energy consumption. In other words, as the economy expands there has been a parallel expansion in the amount of energy used, and the ratio between energy use and GDP is called the "energy intensity." The energy intensity of a country depends on both the overall efficiency of energy use, as well as on the economic structure, and geography of that country. It can be expected that countries with a cold climate, and those with significant energy-intensive resource industries, will in general have a greater energy intensity than those in warmer regions, and those with an economy heavily weighted to the service sector. For example, although Canada is a country with a relatively high energy efficiency, it has one of the highest energy intensities in the world. This strong linkage between overall economic growth and primary energy consumption has been one of the major factors affecting world energy consumption, and the associated production of greenhouse gases. However, in recent years,

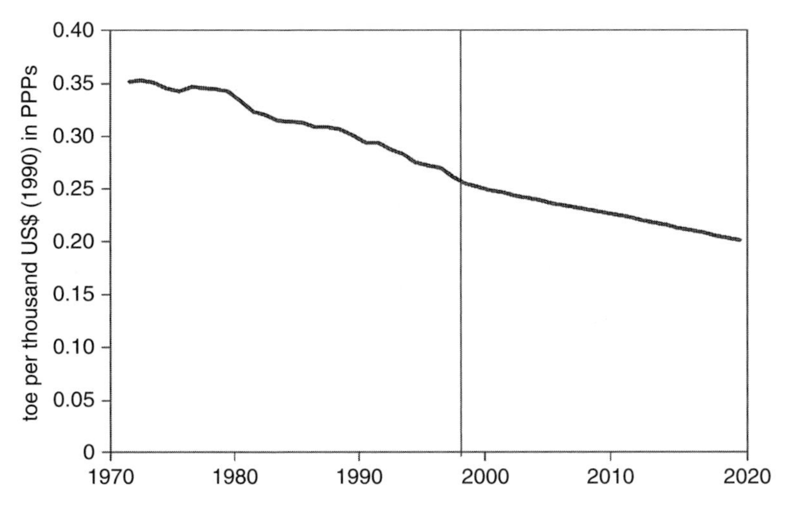

Figure 4.5 World primary energy intensity. *Source*: International Energy Agency *World Energy Outlook 2000*.

with the introduction of many energy efficiency measures, both in the conversion of primary energy into secondary energy carriers, such as electricity, and with improvements in demand-side management to reduce consumption in buildings, for example, this linkage has been weakened. This can be seen in Figure 4.5 (IEA, 2000), which shows the tonnes of oil equivalent (toe) used worldwide per $1000 in PPPs (the PPP, or "purchasing power parity" equalizes the purchasing power of different currencies relative to the US dollar). It can be seen that over the 30-year period from 1970 until 2000 there has been a steady decrease of this ratio from 0.35 in 1970 to approximately 0.25 in 2000, or a 29% reduction. It is unclear if this type of steady reduction can be maintained, although the International Energy Agency (IEA) has predicted that the energy intensity will continue to fall to about 0.20 in 2020, as shown in Figure 4.5. If such a continuous reduction in energy intensity can be maintained, particularly in the rapidly growing emerging economies, it will certainly help to lessen the impact of economic growth on energy consumption and also, therefore, on greenhouse gas emissions.

In terms of primary energy demand by region, estimates by the IEA for the years 2010 and 2030 are shown in Figure 4.6. The estimated 35% increase in demand over this period represents an annual compound growth rate of approximately 1.5%. Although continuous modest growth is predicted for the OECD countries, the share of total

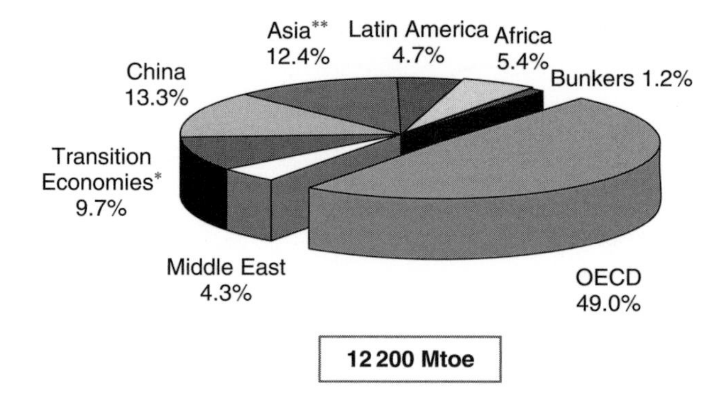

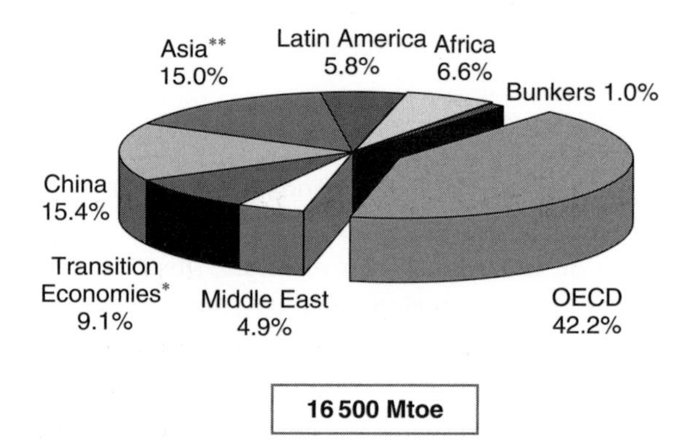

*Includes Former USSR and Non-OECD Europe. **Asia excludes China.

Figure 4.6 Projected world primary energy demand by region in 2010 and 2030 (Mtoe). *Source*: Based on figures from the International Energy Agency *Key World Energy Statistics 2004*.

primary energy is expected to fall for this region, while an increase in the share of primary energy demand is shown for China and other Asian countries. This mirrors the strong growth in demand for all forms of energy in recent years in China and India, as was shown in Figure 4.4. The very strong demand in these emerging economies will likely be a dominant theme, at least until the middle of the twenty-first century, and perhaps beyond. This will be driven by the rapid increase in industrialization in these countries, as well as the desire for large segments of the population to have access to improved transportation facilities, whether by public transport or widespread adoption of personal motor vehicles, as in most of the developed world.

BIBLIOGRAPHY

International Energy Agency (2000). *World Energy Outlook.*
International Energy Agency (2004). *Key World Energy Statistics.*
US Department of Energy. Energy Information Agency (2005). *http://www.eia.doe.gov/*

5

World energy supply

5.1 WORLD ENERGY SOURCES

In Chapter 2 we noted that there are only three sources, or categories, of primary energy; fossil fuels, renewable energy, and nuclear power. The global consumption of energy, broken down into these categories, is shown in Figure 5.1a. It can be seen that almost 80% of all of our primary energy needs are supplied from fossil fuels. The distribution of energy supply by source is further broken down in Figure 5.1b, which shows the largest fossil fuel component of the overall global supply to be oil, followed by coal, and finally natural gas. In the renewable energy category by far the largest component is for combustible renewables and waste, which includes wood-waste and "black liquor" used to fuel boilers in the pulp and paper industry, for example, as well as other combustible biofuels such as firewood gathered by hand in developing countries. The remainder of the renewable energy supplied in 2002 consisted of hydroelectric power, accounting for 2.2% of global demand, while only about 0.5% of total energy demand (shown as "Other" in Figure 5.1b) was supplied from wind, solar, and geothermal power. These figures illustrate the overwhelming reliance that the world places on fossil fuels to satisfy our energy needs. Although crude oil is the largest source of energy, and is used primarily to provide fuel for transportation, we also consume large quantities of natural gas and coal, mainly to provide heat and to generate electricity.

Electricity is not a primary energy source, but is rather an "energy carrier," as we have seen in Chapter 2. However, electricity production is a major consumer of primary energy, and most of the world's consumption of coal, as well as some of the natural gas and all of the nuclear and hydroelectric energy, is used to produce electricity. Although some oil is also used to produce electricity, it is usually

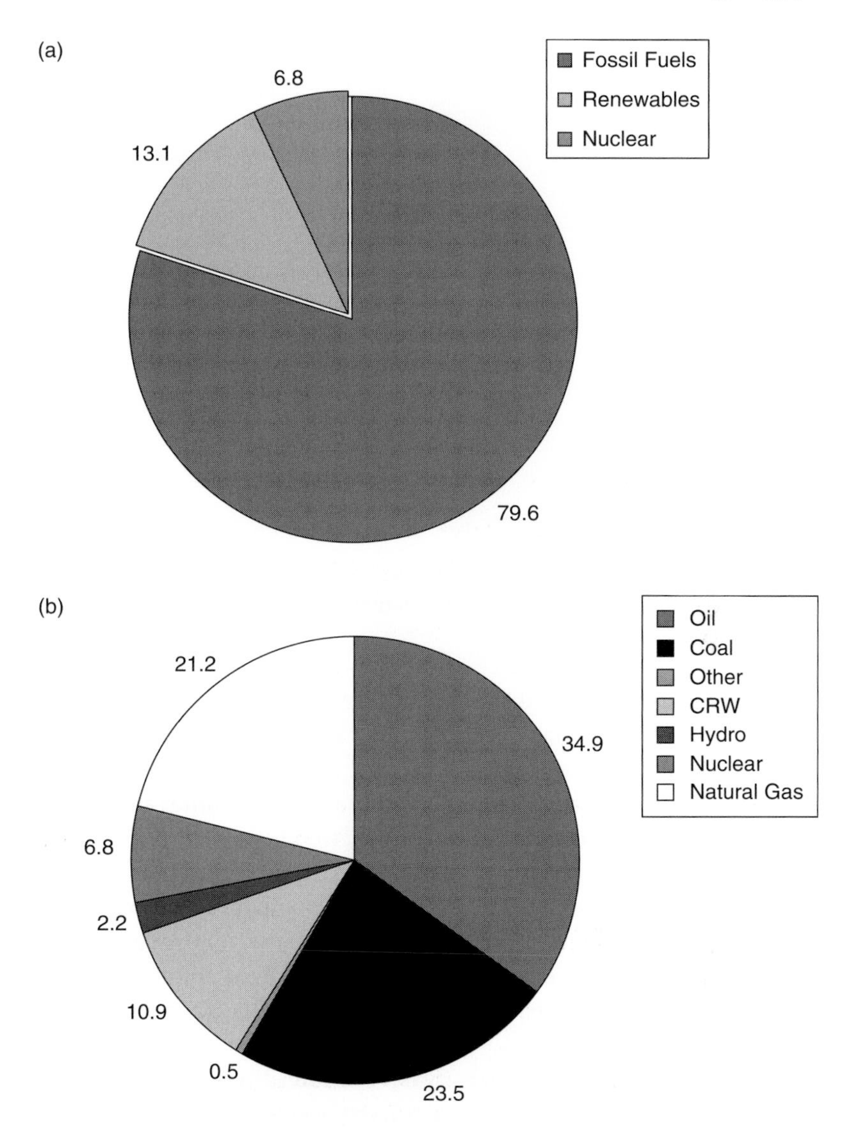

Figure 5.1 World primary energy consumption by source – 2002. CRW = Combustible renewables and waste; Other – Geothermal, Solar, Wind, etc. *Source*: Based on figures from the Energy Information Administration *International Energy Annual 2002*.

limited to small plants in remote areas, or where other sources are simply not available. Figure 5.2 shows the "fuel" share for worldwide electricity production in 2001, as reported by the US Department of Energy (DOE). Nearly two-thirds of all electricity production is from

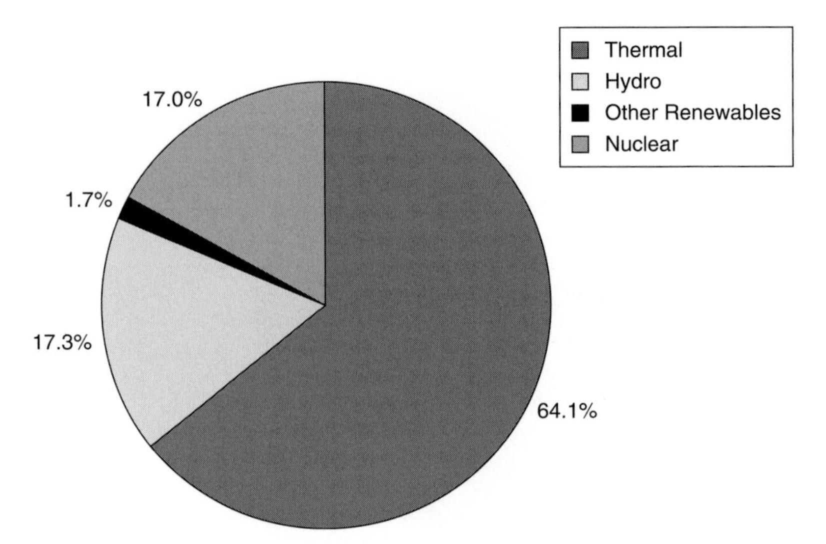

Figure 5.2 World electricity generation by source – 2001. *Source*: Based on figures from the Energy Information Administration *International Energy Annual 2002*.

conventional fossil-fueled thermal power plants, while hydroelectric plants and nuclear plants each supply approximately 17% of total electricity demand. The remaining share, just under 2% of total world electricity production, is obtained from a range of other renewable sources, such as geothermal, solar, wind, wood, and waste. Of course this mix varies significantly from region to region, depending on local availability and cost of the various primary energy sources, and the state of economic development. Nuclear power, for example, naturally tends to be concentrated in the industrialized countries, while hydroelectric power generation is constrained by geographical considerations.

Even though energy intensity has been steadily declining over the last 30 years, as was seen in Figure 4.5, the overall growth in economic activity has resulted in a steady increase in worldwide consumption of all forms of primary energy. The market share of world primary energy supply by source, in millions of tonnes of oil equivalent (Mtoe), is shown in Figure 5.3 for the years 1973 and 2003. The growth in total energy supply, over this period has been nearly 70%, representing an annual compound growth rate of 1.8% (IEA, 2004). Although the growth in market share for some primary energy sources, such as coal and combustible renewables and waste, has been relatively modest in recent years, for others such as natural gas, there has been a steady increase

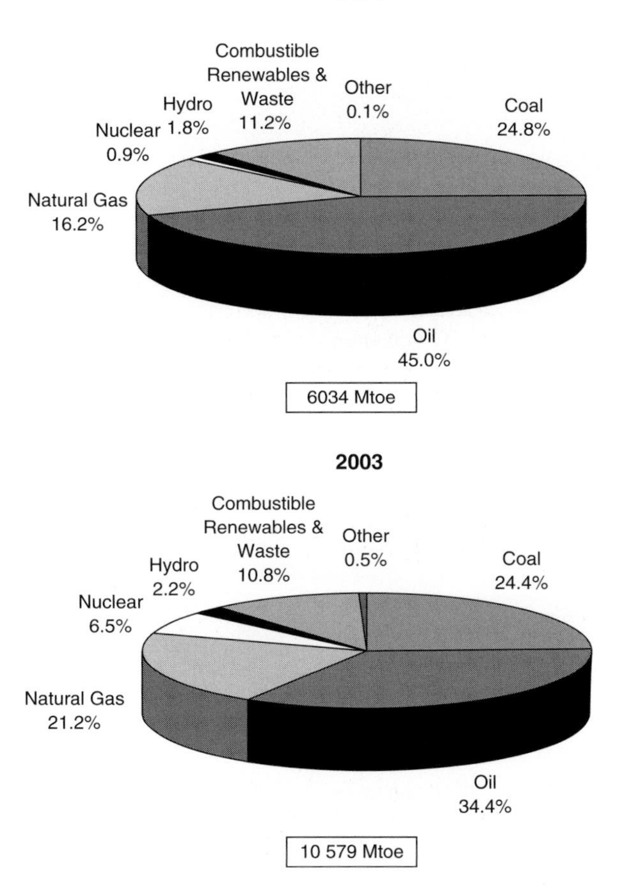

Figure 5.3 World primary energy consumption by source 1973–2002 (Mtoe). *Source*: Based on figures from the International Energy Agency *Key World Energy Statistics 2004*.

in market share. This is particularly the case for natural gas, where in many developed countries, such as the USA and the UK, there has been widespread substitution of natural gas in place of coal for electrical power generation. This has been driven in part by the desire to reduce pollution from burning coal, and in part by the relatively low cost and high efficiency of natural gas-fired combined-cycle gas turbine power generation. Another factor driving this "dash for gas," as it has been called, was the widespread availability and low price of natural gas in many markets. However, quite recently there has been strong pressure

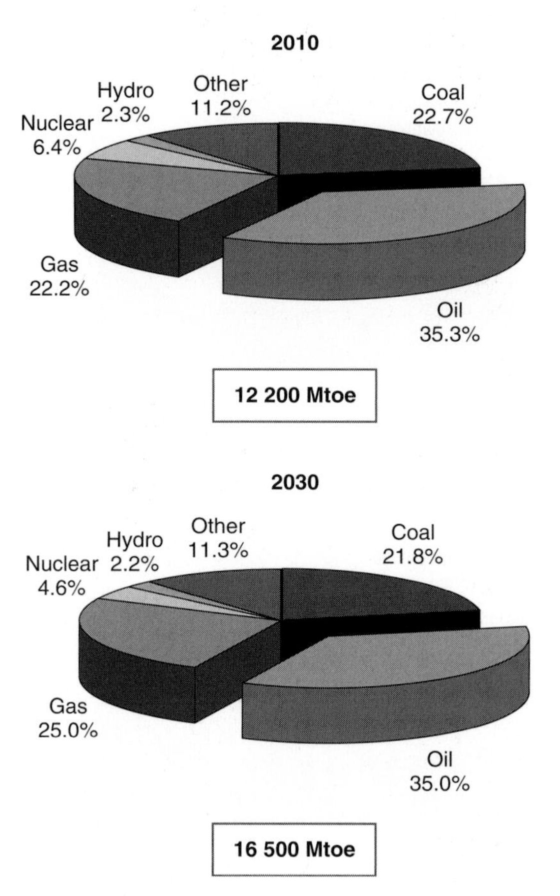

Figure 5.4 Projected world energy consumption by source to 2030 (Mtoe). *Source*: Based on figures from the International Energy Agency *Key World Energy Statistics 2004*.

on gas supplies, with consequent increases in gas prices which has resulted in some concern about the wisdom of such widespread fuel-switching. Although natural gas is a relatively abundant fuel around the globe, it is often located a long way from markets, and it is not as easily transported as is oil. This is particularly true, of course, for transportation by sea, although there is increasing interest in the expansion of seaborne shipment of liquefied natural gas (LNG) from regions with large surpluses of natural gas, such as the Middle East, to regions with high demand, such as the USA and Japan. The market share taken by oil over this period has actually gone down, likely reflecting a switch from the use of oil for electricity generation to other fuels such as coal and natural gas, and improvements in the efficiency of vehicle engines. There has also been significant growth in the supply of nuclear power for electrical generation in the 1970s and early 1980s, although this growth has now slowed dramatically due to concerns from the general public about the safety and long-term environmental effects of nuclear power. As pressure continues to build to limit the production of CO_2, however, there may well be a return to more widespread deployment of nuclear power. We shall discuss this issue in more depth in Chapter 8.

The IEA has also provided some estimates of the increase in demand for all forms of primary energy for the 20-year period from 2010 to 2030, as shown in Figure 5.4. The overall increase in energy demand for this 20-year period is approximately 35%, going from 12.2 Gtoe in 2010 to 16.5 Gtoe in 2030, reflecting an assumed annual compound growth rate of 1.5%, somewhat lower than experienced in the previous 30 years. The growth in demand has been assumed to be higher for some energy sources than for others. For example, the predictions show a decrease in market share for hydroelectric power, primarily due to the relatively small potential for large new sources of hydroelectric power near to major load centers, and also a large decrease in market share for nuclear power due to the often hostile perception of this form of energy by the general public. The largest market share is predicted to be still for oil and natural gas, with transportation driving the demand for oil, and electricity generation and space heating of homes, offices, and factories driving the demand for natural gas.

5.2 FOSSIL FUEL RESOURCES

As we have seen, fossil fuels are the predominant primary sources of energy, providing just under 80% of all global energy requirements in 2002. Since crude oil, natural gas, and coal, are all non-renewable in nature, the question of how long we can continue to rely on them as primary energy sources naturally arises. Although the price of fossil fuels, particularly oil and natural gas, has been at record levels in recent years, the demand for these critical energy resources shows little sign of moderating, and is in fact increasing year by year. Although coal is widely available in many regions of the world, crude oil and natural gas are very unevenly distributed, with large resources of oil concentrated particularly in the Middle East, and very large gas resources in both the Middle East and Russia. The term "resources" is a fairly general one, which usually includes the "proved recoverable reserves," as well as an estimate of what may be recoverable in the

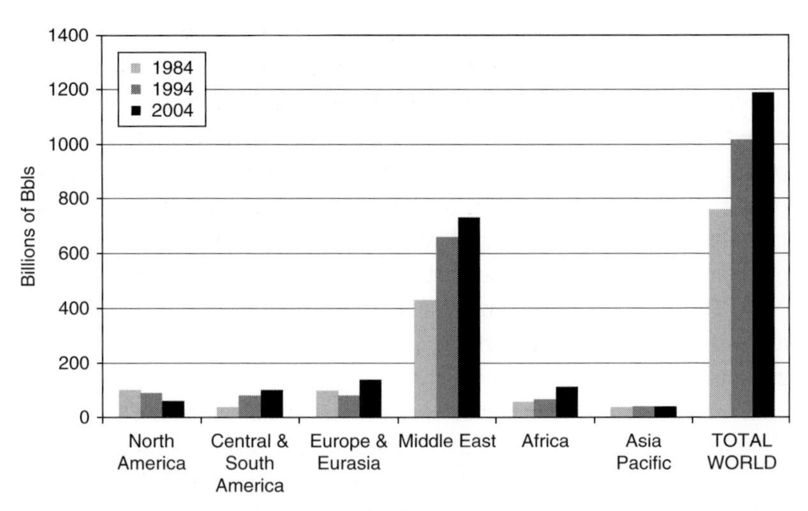

Figure 5.5 Proved recoverable oil reserves – 1984–2004. *Source*: Based on figures from the BP Statistical Review of World Energy June 2005.

future in light of new technological developments for deep-drilling, for example, or a new economic situation which makes currently non-economic resources worth recovering. A more precise definition of "proved recoverable reserves" has been given by the World Energy Council as: "the tonnage within the proved amount in place that can be recovered in the future under present and expected economic conditions with existing available technology." Figure 5.5 (taken from British Petroleum, 2005) shows the proved recoverable reserves of crude oil, in billions of barrels, for the years 1984, 1994, and 2004 (one barrel, or Bbl, is 42 US gallons or 35 Imperial gallons). It is interesting to note that the total reserves in the biggest producing regions, and in the world as a whole, have gone up significantly over this period, although as we shall see this has been matched by increasing levels of consumption.

A more detailed breakdown by country of recoverable reserves in the year 2000 is provided in Table 5.1 (World Energy Council, 2005), in both millions of tonnes of oil equivalent (Mtoe), and millions of barrels (mmBbls). Upon entering the twenty-first century the global total of proved recoverable conventional oil reserves was approximately one trillion (a million, million) barrels. It can be seen that oil reserves are clearly dominated by the Middle East, which accounts for two-thirds of the world's proved reserves of oil. Although the Middle East has a relatively small population and only limited industrial activity,

Table 5.1. *World proved recoverable reserves of oil and natural gas liquids in 2000*

	Mtoe	mmBbls
Africa		
Algeria	1235	10 040
Egypt	529	4150
Libya	3892	29 500
Nigeria	3000	22 500
Other	1466	10 777
Africa - total	10 122	76 967
North America		
Canada	779	6402
Mexico	3858	28 260
USA	3728	29 671
Other	208	1417
North America - total	8573	65 750
South America		
Argentina	429	3054
Brazil	1172	8415
Venuzuela	11 048	76 785
Other	721	5115
South America - total	13 370	93 369
Asia		
China	4793	35 085
India	645	4799
Indonesia	707	5203
Kazakhstan	742	5417
Malaysia	513	3900
Other	783	5832
Asia - total	8183	60 236
Europe		
Russia	6654	48 573
Norway	1510	11 669
UK	665	5003
Other	666	5058
Europe - total	9495	70 303
Middle East		
Iran	12 667	93 100
Iraq	15 141	112 500

Table 5.1. *(cont.)*

	Mtoe	mmBbls
Kuwait	13 310	96 500
Saudi Arabia	35 983	263 500
United Arab Emirates	12 915	98 100
Other	2226	16 553
Middle East – total	92 242	680 253
Oceania		
Australia	445	3848
Other	57	439
Oceania	502	4287
WORLD TOTAL	142 487	1 051 165

Notes:
Oil sands and oil shale reserves are not included.
Data in barrels have been converted at average specific factors for crude oil and NGLs respectively, for each country.
Mtoe, millions of tonnes of oil equivalent; mmBbls, millions of barrels.
Source: World Energy Council.

the region dominates the Organization of Petroleum Exporting Countries, and thus the world oil markets.

A somewhat similar picture emerges for natural gas, as can be seen in Figure 5.6 (British Petroleum, 2005), which shows the proved recoverable reserves of natural gas in trillions of cubic meters for the years 1984, 1994, and 2004. Again, there has been a substantial increase in total world reserves over this 20-year period. The over-riding dominance of the Middle East in terms of crude oil reserves is tempered in the case of natural gas by the very substantial resources in the Russian Federation, which is included in the reserves for Europe and Eurasia. Also, since Russia is closer to the large potential markets for gas in western Europe, and since it is much easier to transport gas by pipeline than by Liquefied Natural Gas, or LNG, carriers, Russia will likely end up as a larger supplier of natural gas than the Middle East, even though both regions have comparable reserves.

The oil and gas business is dominated by a relatively small group of multinational energy companies, and these companies are naturally interested in maintaining their businesses, and strive to meet the increasing demand. They do this by aggressive exploration for new sources, at the same time as they produce oil and gas from their

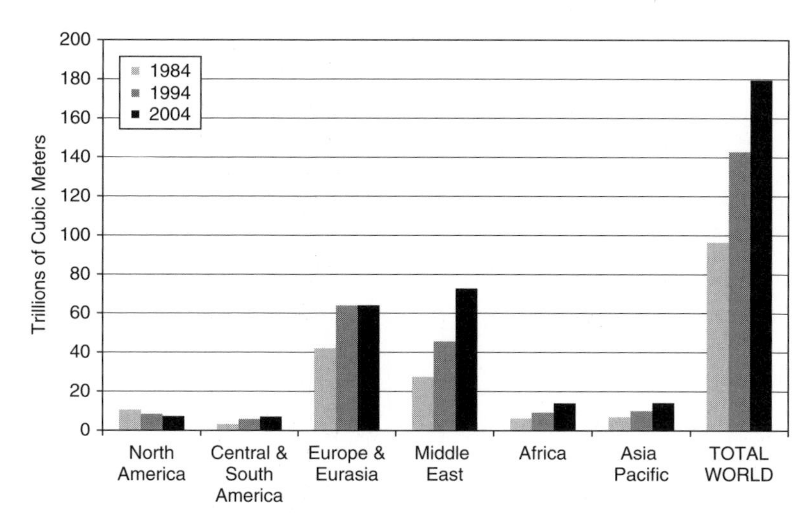

Figure 5.6 Proved recoverable natural gas reserves – 1999. *Source*: Based on figures from the BP Statistical Review of World Energy June 2005.

existing fields. Each company attempts to increase its stock of resources in order to be able to maintain a viable business and to match global demand for their products well into the future. The way in which these companies track their progress is to calculate the "reserves-to-production ratio," R/P, at the end of each year by dividing their stock of "proved reserves" by their annual production. These are tracked by all companies for their own purposes, and most also keep track of all production and exploration worldwide in order to judge their own competitiveness.

Figures 5.7 to 5.9 show the worldwide R/P data compiled by British Petroleum (British Petroleum, 2005) for oil, gas, and coal. The global R/P ratio for oil and gas is shown in Figures 5.7 and 5.8 for every 2 years from 1980 to 2004. It can be seen that in 1980 the global ratio of reserves to production for oil was just under 30, meaning that if the then current rate of production remained constant the proven reserves of oil would be depleted in 30 years' time. The ratio steadily increased through the 1980s, reaching a peak of nearly 45 by the end of the decade. This increase in R/P ratios, even in the face of steadily increasing consumption, is due to the extensive exploration and field development work carried out by oil companies worldwide. For the last decade, however, the R/P ratio for oil has leveled out at around 40, and recent news reports by several oil companies have indicated that

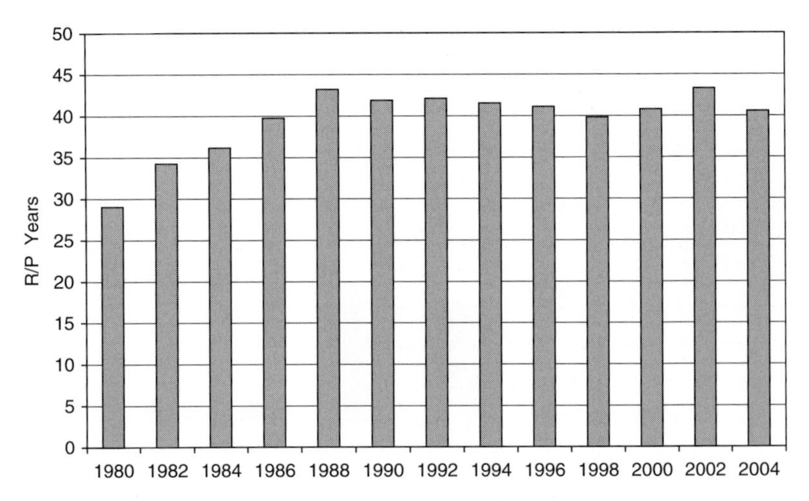

Figure 5.7 Oil reserves-to-production ratio – 1980–2004. *Source*: Based on figures from the BP Statistical Review of World Energy June 2005.

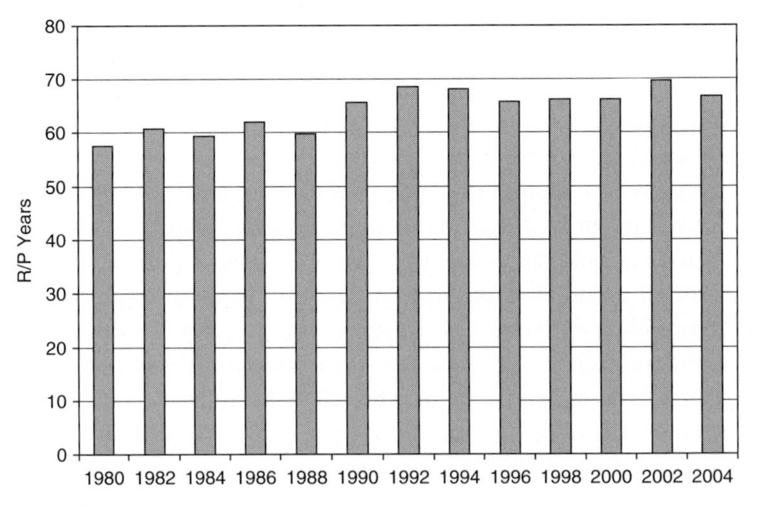

Figure 5.8 Natural gas reserves-to-production ratio – 1980–2004. *Source*: Based on figures from the BP Statistical Review of World Energy June 2005.

they are having to be much more aggressive in their exploration activities in order to maintain the ratio at this level. In the Middle East, by far the largest source of crude oil, the R/P ratio is over 80, while in the major oil-consuming regions of North America and Europe and Eurasia, the ratio is now well under 20.

Similar data for the R/P ratio for natural gas are shown in Figure 5.8. The trends, both globally and by region, are similar to those for oil, although the ratio is significantly higher, and there has been less growth over the last two decades compared with that for oil. There has been a large increase in gas use over this period, however, with gas consumption increasing by 75% over the period from 1984 to 2004. Much of this increased gas supply has been used to generate electricity with efficient combined-cycle power stations, which have replaced coal-fired stations in many parts of the world. The global R/P ratio for natural gas was nearly 70 in 2002, but of course it also varies widely by major producing region, as it does for oil. The ratio for the Middle East is very large, at nearly 280, but this in large part reflects the relatively small demand for gas in the region, and the difficulty of transporting natural gas over long distances compared with the relative ease of transporting oil by sea. There is little doubt that the transportation of gas, both by pipeline and by sea in the form of LNG, will increase in the coming decades in order to exploit the very large reserves of natural gas that are "trapped" in regions such as the Middle East and Africa.

Increasing demand and the resulting high prices for natural gas in recent years has also led to renewed interest in the development of "unconventional gas" sources, such as "tight gas" and coal-bed methane. Tight gas formations are quite widespread, and consist of natural gas trapped in low-permeability porous rock or sand formations. New techniques are being developed to enable this gas to be recovered, including techniques such as hydraulic fracturing and water injection that are similar to those used for enhanced oil recovery. Coal-bed methane, as the name implies, is methane gas formed and then trapped within coal formations, and we shall examine this in more detail in the next chapter. Both of these unconventional sources may result in a substantial increase in proven gas reserves once the new production techniques are fully established.

The R/P ratio for coal in 2003 is shown for the world as a whole in Figure 5.9, and also by major producing region. Because coal has been a relatively low-value fuel, compared with oil and natural gas, and therefore less attractive to ship, there is a greater balance between coal production and consumption in the major coal-producing regions. Most areas of the world have R/P ratios for coal well over 200, with the important exception of Asia, in which there is a rapidly increasing demand for coal for electricity production, particularly in China and India. The overall global R/P ratio for coal, however, is still nearly 200,

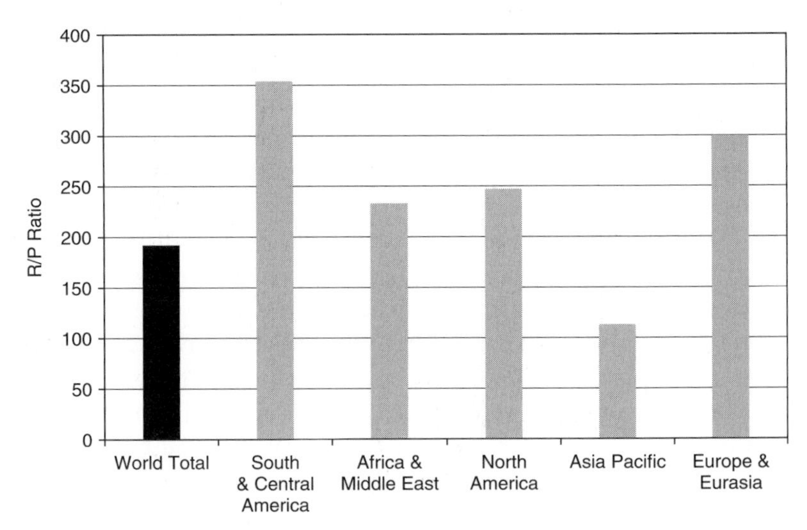

Figure 5.9 Coal reserves-to-production ratio – 2003. *Source*: Based on figures from the BP Statistical Review of World Energy June 2003.

despite the fact that there is little exploration for new sources of coal, at least compared with that for oil and gas. With more aggressive exploration, it is quite likely that the global R/P ratio for coal would increase significantly above the 200 level.

In this chapter we have seen that world energy supply in the twenty-first century continues to be dominated by fossil fuels, accounting for nearly 80% of total global energy supply. Oil is the most widely used fossil fuel, supplying just over one-third of our total energy needs, principally because of its widespread use as a transportation fuel to power motor vehicles, and the ease with which refined petroleum products can be stored and transported. Oil is also a non-renewable resource, and has the lowest reserves-to-production ratio, at around 40, of all the fossil fuel sources. Although an R/P ratio of 40 may seem to be quite high, there is little doubt that continuing increases in energy demand, and more difficulty in finding and developing new sources of crude oil, will highlight the need to develop alternative sources of energy towards the middle of this century and into the next.

5.3 THE GLOBAL DEMAND–SUPPLY BALANCE

As in any economic system, energy demand and supply are balanced by establishing appropriate commodity pricing in world markets. This

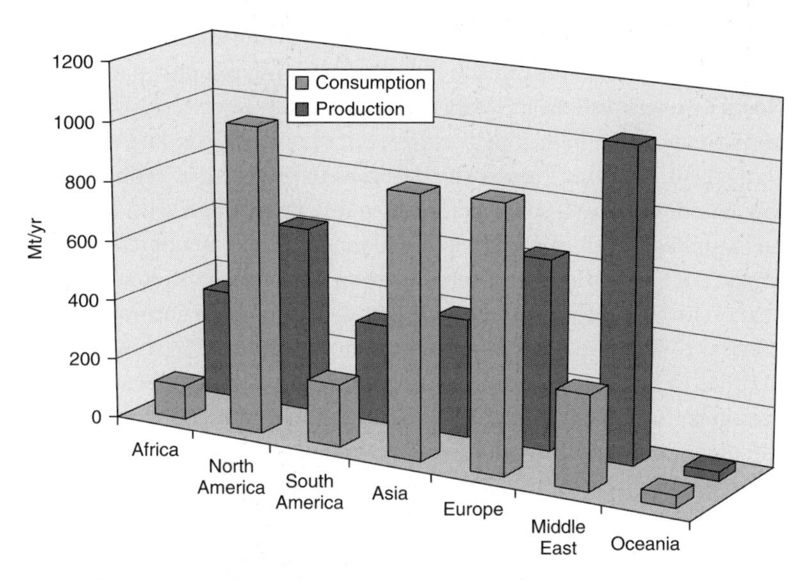

Figure 5.10 Oil consumption and production – 1999. *Source*: Based on figures from the World Energy Council *Survey of Energy Resources Report 2001*.

can sometimes be done on a regional, or national basis, if energy supplies are readily available locally, or through international trade if they are not. As we have seen, energy supply is dominated by fossil fuels, and these are not evenly distributed around the world. This is particularly true for crude oil, which accounts for just over a third of total global energy demand, and so there is a large international trading and transportation system to move oil supplies from the principal producing regions, like the Middle East, to the principal consuming regions, like North America, Europe, and Asia. Figure 5.10 (World Energy Council, 2005) shows both the consumption of oil, in millions of tonnes per year, and oil production, for the major regions of the world in 1999. Although there is substantial oil production capacity still in the major consuming regions of North America, Europe, and Asia, there isn't enough to satisfy the total demand. The major oil-producing region in the world is the Middle East, which has a relatively small population, and therefore low demand, so that oil is transported from there to the major consuming regions. Although Africa and South America produce much less oil than the Middle East, they also have relatively low demand, and so they are also significant suppliers of oil to the larger consuming regions.

When examining Figure 5.10, the disparity between production and consumption of oil in the major consuming regions, particularly North America and Asia, is very clear. The balance between production and consumption in Europe is closer, but that is likely to be a relatively short-term situation, since the reserves-to-production ratio is quite low, and Europe will soon be forced to rely more heavily on imported oil supplies. With such large imbalances in the major consuming regions of the world, a very large portion of their wealth is spent each year on importing oil from other regions. The question naturally arises, then, as to why the demand for oil can not be reduced in these regions in order to bring demand and supply more closely together. The main reason is that the demand for gasoline and fuel oil is what economists call "price inelastic," in other words a higher price results in very little change in demand for fuel. This means that higher taxes on automotive fuels, for example, will not have much impact on the distance that people will drive, or on the size of the vehicle they buy, and therefore on fuel consumption. This can be clearly seen by looking at Europe where the price of gasoline and diesel fuel is more than twice the price in the USA. Although European vehicles tend to be smaller, and therefore more fuel efficient, there is little difference in the per capita number of cars on the road, or in the number of miles driven each year in Europe compared with the USA.

The consumption and production of natural gas around the world, in billions of cubic feet, or Bcf, per year, is shown in Figure 5.11 (World Energy Council, 2005). It can be seen that in this case there is a much closer correlation between consumption and production than is the case for oil. This is in part because natural gas supplies are more evenly distributed around the world, but also largely due to the difficulty of transporting natural gas over large distances, particularly by sea. There is, therefore, only limited inter-regional trading in natural gas, although Asia imports significant quantities by sea in the form of LNG, and Europe is increasingly importing gas from Africa and from countries that made up the former Soviet Union. As gas supplies come under more pressure in the major consuming regions, particularly the rapidly developing countries in Asia like China and India, we can expect to see an increase in the transportation of gas by sea in the form of LNG. This likely will be true as well for Japan, which relies almost entirely on imported sources of fossil fuels, and where gas may be substituted more widely for oil in applications where that is feasible.

The consumption and production of coal is also quite closely balanced, as shown in Figure 5.12 (World Energy Council, 2005). This

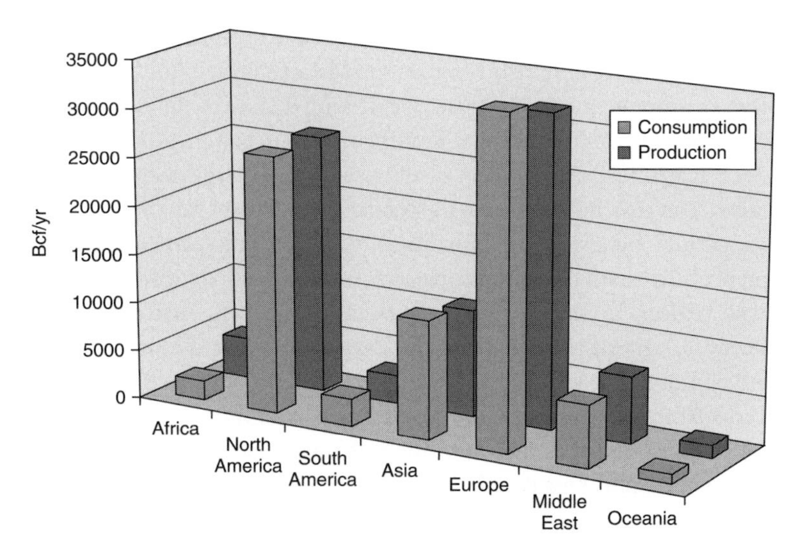

Figure 5.11 Natural gas consumption and production – 1999.
Source: Based on figures from the World Energy Council *Survey of Energy Resources Report 2001*.

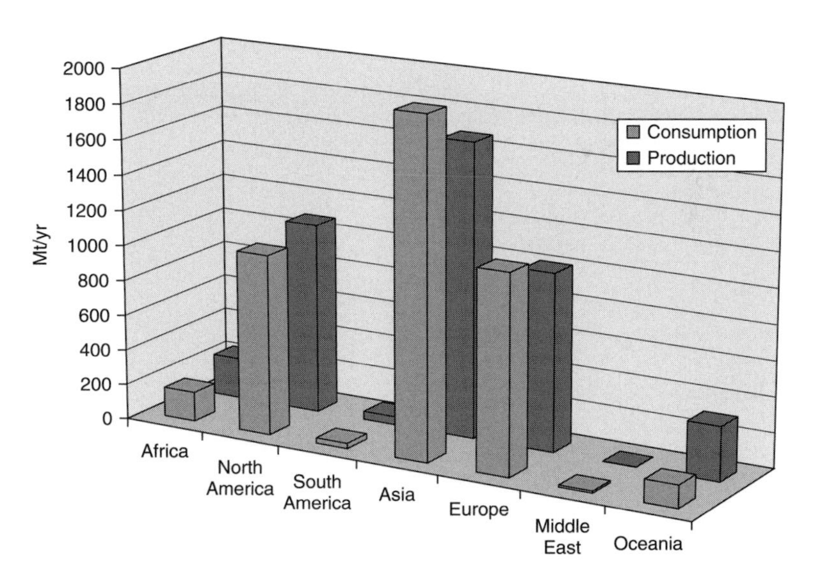

Figure 5.12 Coal consumption and production – 1999. *Source*: Based on figures from the World Energy Council *Survey of Energy Resources Report 2001*.

is because coal is much more widely available around the world than either oil or gas, and also because coal is a relatively low-value fuel, often it is not economic to transport it long distances. In most regions of the world, therefore, local coal production matches the demand, which is primarily for "steam coal," used to fuel large thermal power-plants, and to a lesser extent for "coking coal" used for iron and steel production. There is some imbalance, however, between consumption and production of coal in Europe and Asia. The imbalance in supply in these regions is largely made up by imports from Africa and from "Oceania," in particular from Australia. In Europe the imbalance between demand and supply is due in large measure to the high cost of coal from local deep underground mines, and the relatively low cost of coal from overseas. In Asia, this imbalance is primarily due to the very high demand for coal in China, which has a very rapidly growing demand for steam coal to produce electricity, and due to the relatively short distance from Australia.

BIBLIOGRAPHY

British Petroleum (2005). *Energy in Focus: BP Statistical Review of World Energy June 2005.*

International Energy Agency (2004). *Key World Energy Statistics – 2004.* Paris, France: IEA.

International Energy Agency (2005). *Key World Energy Statistics–2005.* Paris, France: IEA.

US Department of Energy (2005). Energy Information Agency. *http://www.eia.doe.gov/*

World Energy Council (2005). *http://www.worldenergy.org/wec-geis/default.asp*

Part III New and sustainable energy sources

6

Non-conventional fossil fuels

6.1 NEW SOURCES OF OIL AND GAS

We have seen in the previous chapter that there will be considerable pressure on conventional fossil fuel reserves over the next few decades. Demand for oil in particular will experience substantial annual growth, and it will be difficult to maintain the recent historical reserves-to-production ratio of around 40. There is a need, therefore, to develop new or "non-conventional" sources of fossil fuels to supplement the traditional crude oil supplies. These will likely be needed until at least the end of the twenty-first century, when extensive supplies of truly renewable, or sustainable, primary energy should be available in sufficient quantities to satisfy most global energy demand. In the near-term these "new" sources of fossil fuels include the unlocking of "synthetic oil" from the extensive oil sands and oil shale deposits found in many parts of the world, and the extraction of natural gas from unused coal seams, known as "coal-bed methane." In the longer term the use of fossil fuels in a much more environmentally benign way may be prolonged by accessing the extensive global coal supplies using so-called "clean coal" technologies, or even by accessing the extensive methane hydrate resources to be found in the deep ocean. If carbon mitigation, in the form of CO_2 capture and storage, also known as "carbon sequestration," is proven to be technically and economically viable, then we may still be using fossil fuels well into the twenty-second century. In this chapter we will briefly review the current state of development of these new or non-conventional sources of fossil fuels.

Canada has vast resources of both oil sands and heavy oil, mainly located in the Athabasca region of Northern Alberta. The ultimate resource of bitumen trapped in the sands in this region alone has been estimated by the Alberta Energy and Utilities Board (AEUB) to be the equivalent of 2.5 trillion barrels of conventional oil. As with most oil deposits, it will not be economically or technically feasible to recover all of these resources. The AEUB estimates, however, that some 315 billion barrels of synthetic crude oil, comparable to the proven reserves of Saudi Arabia, may be ultimately recoverable. Of these resources, nearly 175 billion barrels are classified as proven reserves, recoverable with current production techniques. This represents a significant increase in the total proven reserves of oil that will be available worldwide, currently estimated by the World Energy Council to be about 1 trillion barrels. The heavy oil, or bitumen, is trapped in a mixture of sand, water, and clay, which until recently provided a significant challenge for extraction. However, over the last 40 years research and development work has resulted in two quite different extraction techniques to unlock this huge resource of synthetic crude oil. Several large plants are now producing a high quality synthetic crude oil from the Athabasca region, and extensive plant expansions are either under way or at an advanced planning stage.

The earliest technique used to extract synthetic crude oil from the extensive Canadian oil sands deposits starts with open-cast mining of the very large surface deposits found in the Athabasca region of Northern Alberta. The bitumen-containing sands are then trucked to a nearby extraction plant where the bitumen is separated from the sand using hot water, and subsequently turned into synthetic crude oil in an upgrading plant, before being sent to a refinery. In this process approximately 75% of the bitumen can be removed from the oil sand, and one barrel of synthetic crude oil is produced for every two tonnes of oil sands that are mined and processed. The first such plant, the "Great Canadian Oil Sands Project," now operated by Suncor Energy, was opened in 1967, and this was followed by a second large mining and extraction operation opened by the Syncrude consortium in 1978. For bitumen located deeper underground, mining is not feasible, and several "in-situ" recovery techniques have been developed. These include a simple "two-pipe" technique, known as Steam Assisted Gravity Drainage (SAGD) in which two parallel horizontal wells are drilled, one above the other, in an oil sand deposit. Steam from the top pipe is injected into the oil sand formation, heating the bitumen and lowering its viscosity so that it drains by gravity into the lower perforated pipe. The bitumen is then pumped to an upgrading plant for conversion to synthetic crude. Other in-situ techniques include Cycle Steam Stimulation (CSS), and Vapor Recovery Extraction (VAPEX).

In the CSS technique steam is first injected into the oil sands from a conventional vertical well during a heating cycle, and this is followed by a pumping cycle to remove the bitumen which has been separated from the sand during the previous cycle. The VAPEX process uses a solvent in place of steam to first separate the bitumen from the sand, followed by a pumping operation similar to that used in the CSS process.

At the present time approximately two-thirds of synthetic crude production is from mining operations, with the remainder from various in-situ processes. Both surface mining and in-situ operations have been expanded, and new operations are also being planned, so that oil sands production now accounts for over 30% of Canada's total oil production. The current oil sands production is approximately one million barrels per day, and with additional plants now either being built, or planned, production is expected to grow to more than 60% of total Western Canadian oil production by 2010. Of course there are environmental issues associated with production of synthetic crude oil on a large scale, including disposal of the "tailings," or waste sand stream, and the use of large quantities of water and natural gas in the process. Natural gas is currently used to generate the steam and heat needed to separate the bitumen from the sand and clay materials, and water is used in the separation process. At the present time the energy in the form of natural gas required to produce one barrel of bitumen, is in the range of 10–20% of the energy content in the resulting synthetic crude oil, depending on the recovery process used. However, new methods are continually being developed and improved to reduce energy consumption, and techniques are also being implemented to increase the recycling of water used in the extraction process. For example, reducing the temperature of the hot water used for extraction from 80 °C to 35 °C in one plant has resulted in a substantial reduction in the quantity of natural gas required, and CO_2 emissions produced. After extraction of the bitumen, the sand and remaining solid materials are returned to the mine-site so that reclamation of the land can be completed. There are currently four major plants operating in the Athabasca oil sands region of Northern Alberta. The first two of these to be built and commercially operate are the Suncor Energy plant, producing some 200 000 Bbls/day of synthetic crude, and the Syncrude Canada operation, also producing around 200 000 Bbls/day, but with an expansion under way to extend capacity to 350 000 Bbls/day. The Athabasca Oil Sands Projects, led by Shell Canada, has a current capacity of 65 000 Bbls/day with plans to increase

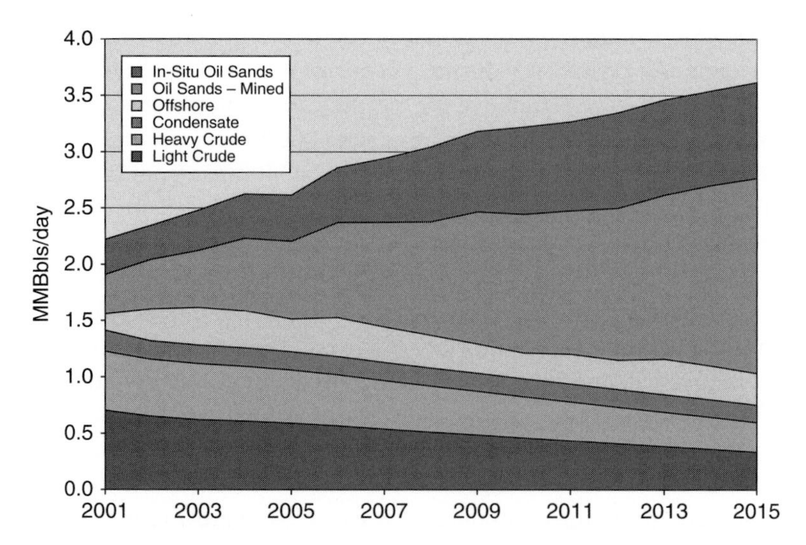

Figure 6.1 Canadian oil production projections to 2015. *Source*: Based on figures from the Canadian Crude Oil Production and Supply Forecast 2006–2020.

this to 150 000 Bbls/day, and finally Imperial Oil operates the Cold Lake facility, with a capacity of just over 100 000 Bbls/day of synthetic crude. Several other plants are also either under construction or in the early planning stage of development.

Total Canadian oil production is shown in Figure 6.1 (Canadian Association of Petroleum Producers, 2005) for the years 2001–2005, with estimates also shown for production out to 2015. It can be seen that over this period total oil production is projected to increase from the current 2.6 million barrels per day to 3.6 million barrels per day by 2015. However, conventional crude oil production from Western Canada is projected to drop from about 1.25 million barrels per day in 2001 to about one-half this level in 2015. Offshore production, which recently came on-stream in Atlantic Canada accounted for some 250 000 barrels per day by 2005, but is not projected to grow much from there. The cost of production of synthetic crude oil from the oil sands has been estimated by the Canadian National Energy Board (2005) to range from a low of $10 per barrel for some in-situ processes to a high of approximately $25 per barrel for mining operations. With world oil prices reaching historic high levels of nearly $70 per barrel in 2005, production from the Canadian oil sands is projected to grow rapidly. The growth in total production of one million barrels per day over the estimate period is entirely due to production of synthetic crude from oil sands, and most of this is from surface mining operations. This is an increase of about 40% over the annual production in 2003, and the growing supply will serve Canada's domestic market as well as some exports to the USA, which is increasingly reliant on imported oil.

Oil shale, which is somewhat similar to oil sands, but with organic matter locked into sedimentary rock formations, is also present around the world in very large quantities. The organic material can be separated from the rock, and converted to synthetic oil using heat and the addition of hydrogen, but it is a much more difficult, and expensive, process than that used for extraction from oil sands. The World Energy Council (2005) has estimated that the USA has more than 60% of the world's total oil shale resources, mainly located in the states of Colorado, Utah, and Wyoming. The total recoverable reserves of oil thought to be obtainable from oil shale in the USA are on the order of 500 billion barrels, or approximately one-half of the total world proven reserves of conventional crude oil. Although several oil companies have taken out oil shale leases in the past, and small pilot plants have been built to test the oil extraction process, the high cost of production, and environmental concerns, has led to all of these attempts being abandoned. Large-scale extraction of synthetic crude oil from oil shale in the USA may one day resume, but it will not likely occur until supply shortages of conventional oil push prices well beyond those in effect today. Oil shale is also found in many other regions of the world, and is sometimes used directly in place of coal in thermal powerplants. Production in Estonia for this purpose, for example, reached some 30 million tonnes per year in 1980, but has since declined to about one-third of that level. If coal prices continue to increase on world markets, then there may be an increase in the use of oil shale for power production, particularly in those countries with limited coal resources.

Another alternative source of fossil fuel is "coal-bed methane," essentially natural gas that has been trapped in coal seams during the formation of the coal. Since coal resources are extensive, and widely distributed around the world, coal-bed methane represents a considerable resource in many regions. Coal deposits provide a very large internal surface area, and because of this they are able to store large quantities of natural gas. According to the US Geological Survey (USGS, 2005), coal deposits are able to store 6–7 times the quantity of methane that can normally be found in conventional porous rock reservoirs of the same size. The resource in the USA has been estimated by the USGS to be around 700 trillion cubic feet. Of that amount, some 100 trillion cubic feet is assumed to be readily recoverable, an amount equal to nearly 60%

of the total US proven reserves of 167 trillion cubic feet. Resources in other parts of the world are not well defined, but are expected to be widely distributed, given the very wide distribution of coal resources. Coal-bed methane is often found in quite shallow coal deposits, and is therefore quite easily produced by drilling wells into these formations. However, unlike in conventional gas fields, the methane gas is usually co-located with water, which must first be pumped out of the deposit, and disposed of in an environmentally acceptable manner. If ways can be found to safely dispose of the large volumes of water co-produced with the gas, then coal-bed methane can be expected to provide a significant addition to conventional natural gas resources around the globe. An added benefit is that the resources are likely to be more widely distributed than conventional natural gas resources, so that gas could be locally produced, rather than having to rely on extensive pipelines or LNG tankers for transportation from other regions. These additional reserves of gas are expected to be more widely explored, and more widely utilized, as conventional natural gas supplies are depleted and gas becomes more expensive on the world market.

The discovery of very large quantities of methane hydrates (methane is the primary constituent of natural gas), complex crystal-like 'clathrate' structures consisting of methane molecules surrounded by ice, on the deep ocean floor and in the arctic permafrost, has led some observers to believe that these will be the long-term future of natural gas supply. In fact, this is such a huge potential energy resource that the USGS has estimated that methane hydrates contain twice the amount of carbon contained in all other fossil fuels on earth! The USGS has said that the deposits on the ocean floor may be up to 13 km deep in some areas, and may also be trapping some methane gas that is not bound up in the crystalline hydrate structures. However, this complex resource is not yet well understood, and practical or economic recovery techniques to unlock the enormous methane resource have yet to be developed. Researchers believe that drilling techniques similar to those used to produce conventional natural gas reserves may also be applicable to methane hydrates, but much more work needs to be done before commercially viable production methods can be demonstrated.

6.2 CLEAN COAL PROCESSES

Coal is widely utilized around the world, primarily for the generation of electricity and the reduction of iron ore in steel mills. Because coal resources are very large, and coal is widely available in many regions, there is considerable interest in the expansion of coal use. Conventional combustion of coal, however, carries with it significant environmental penalties, including emissions of NO_x, SO_x and particulates, all of which can cause significant health effects. Also, because of the high carbon-to-hydrogen ratio, coal produces much greater quantities of CO_2 per unit energy output than do other fossil fuels. This has led to significant research on so-called "clean coal" technology, which seeks to reduce the environmental effects of coal utilization. Since coal resources are much greater than both oil and gas resources, and are widely distributed around the world, there is also a large research effort aimed at converting coal into more transportable and "user friendly" resources in the form of coal-derived liquid and gaseous fuels. In the future, when conventional oil and gas supplies become scarce and therefore expensive, there may be a need to rely on coal as a source of hydrocarbon feedstock for the production of chemicals and plastics.

Most coal-fired power stations today utilize "pulverized fuel" (or PF) technology, in which coal is ground into a very fine powder, and then is blown into the boiler firebox where it is burned in suspension. A conventional boiler raises steam at a pressure of around 165 bar (2400 psi) and a temperature of some 565 °C. Under these conditions, the overall thermal efficiency of the powerplant, taken as the ratio of the electrical power generated to the rate of consumption of coal, is just under 40%. The basic laws of thermodynamics show that increased thermal efficiency can be obtained by increasing the average temperature of the steam, and so engineers are always seeking ways to do this. One way that has now been commercialized is to raise the steam pressure to so-called "supercritical" pressures, in which the steam in the boiler never changes phase from liquid to vapor, but remains as a very dense single-phase fluid. The heat in the boiler is then transferred to this high-density fluid at an average temperature which is significantly higher than that achievable in a conventional boiler which must first convert all of the water into vapor at a relatively low temperature, and then superheat it to a higher temperature. These supercritical plants operate at a pressure of about 240 bar (3500 psi), and as a result the thermal efficiency is increased to around 44%. Although this may seem to be a fairly modest achievement compared with the nearly 40% efficiency achieved by conventional power stations, it does represent a valuable 10% reduction in the amount of fuel required to generate a given amount of electricity, and in the volume of CO_2 emissions produced. Over 400 supercritical powerplants are now installed worldwide, and this technology is rapidly becoming the new "standard" for PF coal-fired power stations. Other techniques for utilizing coal directly to generate electrical power with a higher thermal efficiency, and lower emissions, are also being developed, but have not yet achieved widespread commercialization. These include the Pressurized Fluidized Bed process, in which coal is burned in a compact pressurized vessel which offers the possibility of using the hot combustion gases to drive a gas turbine, while at the same time steam is generated using heat-transfer tubes immersed in the "fluidized bed." The benefit of such a system is the efficiency gain made possible by operating both a gas turbine and steam turbine in a combined cycle approach (described below), which could result in thermal efficiency values approaching 50%. Much development work remains to be done, however, in order to ensure that the combustion gases do not damage the gas turbine because of the carry-over of ash particles from the combustion process.

Both coal gasification and coal liquefaction have been extensively studied, and have been used to produce coal-derived liquid and gaseous hydrocarbons in the past. Because of the wide availability of conventional oil and gas resources in the latter half of the twentieth century, however, these have not been widely exploited. This situation may well change as these conventional resources become increasingly scarce and expensive, and coal may once again be widely used for the production of liquid and gaseous fuels. Coal gasification was being used from the early part of the nineteenth century to produce "town gas," primarily a low calorific value mixture of carbon monoxide and hydrogen. To produce this gas a coal gasifier heats coal in the presence of oxygen and steam so that the coal breaks down into a combustible mixture of carbon monoxide and hydrogen, and other by-products. The combustible town gas can then be distributed to homes and industry to be used for gas lighting and to provide space heating and fuel for industrial processes. The generation and distribution of this important fuel was replaced in the twentieth century, however, by the widespread availability of natural gas, which is now extensively used for industrial processes, and space heating in commercial and residential buildings, as well as to generate electricity. However, the basic coal gasification technology originally developed for the production of town gas has been modified and improved for use in advanced "Integrated Gasification Combined Cycle" (IGCC) power generation plants, and to produce "syngas" as a feedstock for the synthesis of liquid fuels in a "gas-to-liquids" (GTL) process.

The IGCC power generation process enables the coal gas to be used as fuel for a gas turbine, which generates electricity and also provides hot exhaust gases to generate steam in a boiler for subsequent use in a conventional steam turbine. This combination of both gas turbine and steam turbine (the "Combined Cycle"), which is similar to that used in modern natural gas combined-cycle powerplants, provides a much higher efficiency than a simple PF coal-fired steam powerplant due to the higher temperatures achieved in the gas turbine. The fact that the coal is first processed into a gas means that the solid material left over after gasification of the coal is mostly contained as ash in the gasifier rather than passing into the atmosphere as particulate matter. With an overall thermal efficiency approaching 50%, rather than just under 40% as in a large pulverized coal-fired powerplant, the combined cycle also results in a significant reduction in the production of CO_2. At the present time only a few prototype plants have been built, primarily to demonstrate the technology and to act as test-beds for new gasification technologies. However, with further development work to reduce the cost of IGCC plants, and increasing costs for natural gas, it is only a matter of time before many future powerplants are high-efficiency, low-emission coal-fired IGCC units. Also, if oxygen is used, rather than air, as the oxidant in a coal gasifier, the volume of flue gas (or exhaust) is greatly reduced due to the absence of nitrogen, which makes up nearly 80% of the air normally used for conventional combustion or gasification. The flue gas then consists mainly of highly concentrated CO_2, making the process of separating CO_2 for the purpose of capture and storage (discussed in the next section) much easier and less expensive. If carbon capture and storage becomes the accepted way to deal with CO_2 emissions from coal-fired electrical power generation, then IGCC systems will be much more attractive than conventional pulverized coal plants. Another benefit of on-going development of the gasifier component of these plants is that the technology can be readily used in the future to produce syngas, either as a replacement for natural gas, or to produce liquid fuels in a gas-to-liquids plant, when required.

In the longer term, underground coal gasification, in which coal is converted into gas "in-situ" without any need for expensive and environmentally contentious mining, may be the preferred way to turn large portions of the coal resource directly into syngas. This technique could also be used to access coal that would normally be inaccessible, because it is too deep, or not economic to produce with conventional mining techniques. Underground gasification is a quite simple "partial oxidation" process in which the coal is heated to release the volatile components by controlled combustion with less air than would be used for complete combustion. In practice, this process would entail two drill-holes, placed into a coal seam some distance apart. Air, or oxygen, would be blown into one end of the seam to partially burn some of the coal. The heat generated would then gasify the remaining coal, and the syngas would be extracted from the second drill-hole and used directly for either electrical power generation or the synthesis of liquid fuels. This is in principle just like the process in a conventional coal gasifier, but in practice it has been much more difficult to control the process because of the large size and variability of most underground coal seams. Underground coal gasification was pioneered in the UK and in Russia in the 1920s and 1930s, and then further developed in the USA in the 1970s. However, due to the relatively low cost of natural gas, and the difficulty in controlling the underground gasification process, it was never pursued to the stage of large-scale commercial production. With further development, however, and increases in natural gas prices, this technique could prove to be a very cost-effective way to release the energy contained in large coal resources not normally considered economically viable.

The world's first coal liquefaction processes were developed in Germany during the 1920s to ensure that the rapidly growing demand for petroleum products could be met in a country with very small reserves of crude oil, but extensive coal resources. Coal hydrogenation and Fischer–Tropsch gas-to-liquids processes were used to synthesize liquid fuels from coal and from syngas produced by coal gasifiers. Nearly two dozen of these coal liquefaction plants were built during the Second World War so that Germany would not be dependent on imported crude oil to supply the needs of its armed forces. The hydrogenation plants, although disrupted near the end of the war by Allied bombing, were able to supply most of the high-quality aviation gasoline needed for the Luftwaffe up to 1944. In addition, the less well-developed Fischer–Tropsch synthesis was able to supply a significant fraction of the diesel fuel and lubricating oil required by the army and for civilian transport. After the war, these plants were not economically competitive with the inexpensive process of refining the increasingly available crude oil supplies, and were abandoned. The SASOL company in South Africa is the only plant in the world to utilize this technology today to produce liquid fuels, including synthetic gasoline and diesel fuel, from coal. The plant was built in the apartheid era when the South African government was no doubt concerned about the possibility of an oil import embargo, which would have crippled the economy in a country with no natural reserves of crude oil. However, this plant now serves as a showcase to the world for this technology, and demonstrates that technology is available today to produce liquid fuels directly from coal. In 2004 the SASOL plant produced just over 5 million tonnes of liquid and gaseous fuels from coal. As conventional oil supplies become scarcer, and more expensive, we may see more such plants producing synthetic liquid fossil fuels.

6.3 CARBON MITIGATION

We have seen that the world depends on fossil fuels for nearly 80% of its total energy needs, and this will continue for many decades until more sustainable sources of energy supply can be developed and expanded to act as replacements. The end result of all uses of fossil energy is the conversion of the carbon in the fuel into CO_2 gas, which is normally released into the atmosphere. As a result, there is a now a great deal of concern about the effect of global emissions of CO_2, which produce some 6 billion tonnes of carbon per year, as we have seen in Chapter 3. Engineers and scientists have therefore begun to look at the prospects for reducing the quantity of CO_2 vented to the atmosphere by using techniques known as "carbon capture and storage," or "carbon sequestration." The development of these techniques is at a very early stage, but could be one way to prolong the continued use of fossil fuels while at the same time reducing the emission of CO_2 into the atmosphere.

The use of CO_2 for enhancing the recovery of oil from conventional wells is already a proven technique. As shown in the schematic of Figure 6.2, the concept of enhanced oil recovery (EOR) using CO_2 is quite straightforward. A supply of pressurized CO_2 is piped to an area some distance away from the well-head, and is then injected into the oil-containing formation. The pressurized gas then acts to push oil which is left in the porous formation towards an existing drill-hole, and is then extracted with the normal well-head pump. The oil is more expensive to recover compared with a conventional producing well, but EOR can significantly extend the life of a partially depleted oil field, and make better use of the production equipment already in place. Although this is not primarily designed to store, or "sequester" CO_2, the production well is usually capped before significant quantities of CO_2 begin to escape from the well-head. In this case, once the EOR process has been completed, a secondary benefit is the "capture" of CO_2 and storage in the depleted oil formation. The use of CO_2 for EOR

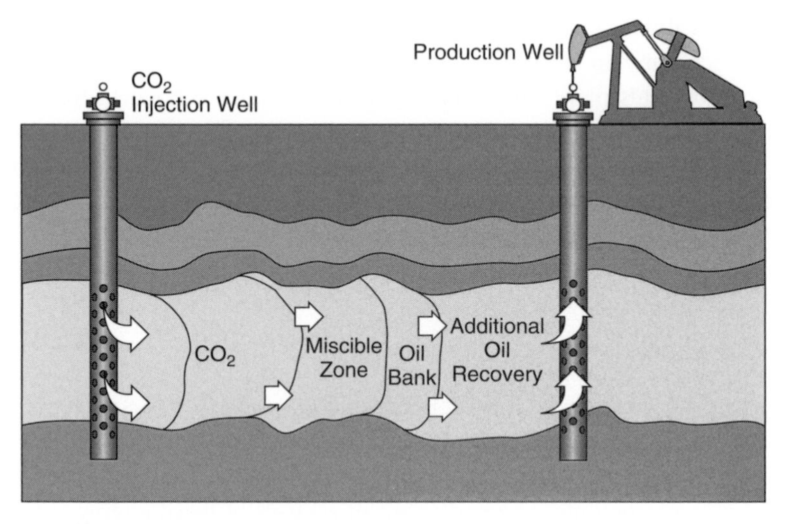

Figure 6.2 Enhanced oil recovery.

has already been applied quite extensively in the USA, and according to the IEA approximately 33 million tonnes of CO_2 per year is currently used at some 75 EOR operations. This process provides the template for one form of carbon capture and storage being proposed by engineers and scientists.

The full range of carbon storage techniques envisaged by researchers is illustrated in Figure 6.3. The emissions of CO_2 from a power station, or some other fossil-fuel consuming process, are first separated from the flue gases in the plant exhaust, and then stored using one of the four different concepts shown. Three of these concepts involve "geologic" storage of CO_2, while the fourth utilizes the deep ocean as the storage medium. In the first case, the gas can be pressurized and piped into a depleted oil or gas reservoir, in the same way as that used in the EOR process. In this case, however, the depleted reservoir is simply used to contain the CO_2 as it no longer contains any economically viable resources of oil or gas. Depleted gas reservoirs are thought to be particularly suitable candidates for this purpose, since they have successfully contained a large gaseous resource for many thousands of years, without significant leakage into the environment. The IEA is currently sponsoring a trial carbon capture and storage project in Canada, in which 5000 tonnes per day of CO_2 is piped from a coal gasification plant across the border in North Dakota, and is then injected into a disused oil field in Weyburn, Saskatchewan. Ongoing monitoring and data collection will be used

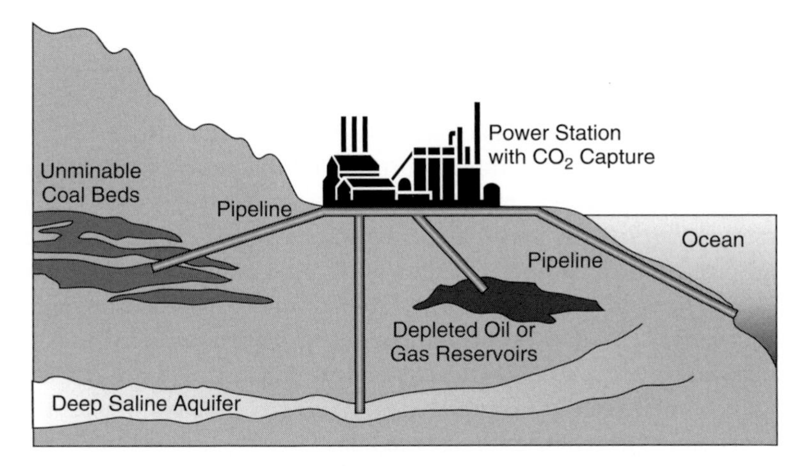

Figure 6.3 Carbon storage concepts.

to see if this is a suitable site for carbon storage, and to provide some valuable information on the costs of such a process.

In a similar way, scientists have speculated that underground aquifers, with large quantities of trapped water, could be suitable repositories for storing large quantities of CO_2, and unused coal seams have also been suggested as possible storage sites. Saline aquifers would normally be sought to store the CO_2 since the water would not be usable for any other purpose. The gas would dissolve in the water, and in time may also react to form solid carbonate materials that would permanently sequester the carbon. In a trial of this technique, nearly a million tonnes per year of CO_2 is being separated from the natural gas being produced in the Norwegian Sleipner field in the North Sea, and is then being piped into a saline aquifer deep under the sea floor. The benefit of using unminable coal seams is that most coal deposits usually contain a large amount of methane, or "coal-bed methane," which is trapped in the porous coal formation. The injection of CO_2 into the coal seam, in conjunction with suitably placed gas extraction wells, can release the methane in a way similar to that used in the EOR technique. The CO_2 then takes the place in the coal bed of the methane, which is a valuable fuel resource once it is recovered.

Finally, a longer-term prospect for the storage of CO_2 is to transport it by ship or pipeline for storage in the deep ocean. Two concepts for deep ocean storage have been proposed, and are being studied under the auspices of the IEA. In the first proposal the CO_2 is injected directly into the ocean at about mid-depth so that it disperses as widely as possible and is dissolved into the seawater. The second proposal is to pump liquid CO_2 to fill depressions on the deep ocean floor. At the pressure and temperature in depths greater than about 3000 m the CO_2 would remain as a liquid with a higher density than water. The liquid CO_2 would then form a stable "lake," and should remain at the bottom of the ocean with only minimal diffusion into the surrounding water. In both cases, however, there is likely to be some "leakage" of CO_2 back to the ocean surface, and this gas will then re-enter the atmosphere. Much research is now being conducted to determine what these leakage rates would be, and to try to estimate the costs of storing large quantities of CO_2 in this way. Although deep ocean storage appears to be attractive because of the very large storage capacity of the ocean, it is unlikely to be seen as a commercially viable technique until after the development of geologic storage methods.

One of the major challenges of implementing carbon capture and storage is the difficulty of efficiently separating CO_2 from the exhaust gas stream. In most coal combustion processes CO_2 accounts for approximately 12% of the total flue gas volume, which consists primarily of CO_2, nitrogen, and water vapor. Nitrogen, which makes up about 79% of the volume of air, does not react in the combustion process, and remains by far the largest component of the flue gases. Separation of CO_2 from the nitrogen and water vapor then becomes challenging due to the very large gas volumes involved and the requirement for large pieces of equipment. There are two main concepts which have been proposed for capturing CO_2 from the combustion process, and these are referred to as "post-combustion" and "pre-combustion" techniques. In the most well-developed post-combustion technique a flue gas "scrubber" is used to separate CO_2 from the rest of the flue gases, using an appropriate solvent. In demonstration plants operated today at near-commercial scale, mono-ethanolamine is used to absorb CO_2 from the flue gas, and the resulting liquid is subsequently heated to release nearly pure CO_2 ready for transportation to a suitable storage site. Because of the very large flue gas volumes, however, the scrubber also needs to be very large, and is quite an expensive piece of equipment. The process of heating the solution to release CO_2 gas also requires an input of energy, thus reducing the overall thermal efficiency of the plant. Other solvents are being investigated to see if the energy requirements can be reduced, as are processes which use porous metallic membranes to directly "filter" the CO_2 from the flue gas streams. These techniques are at an early stage of development, however, and much more research and development work needs to be done before they can be proven to be technically viable and cost-effective.

In the post-combustion concept the volume of flue gas that needs to be processed can be greatly reduced if pure oxygen is used as the oxidant in the combustion process, rather than air. In this way there is little or no nitrogen present, greatly reducing the volume of flue gases which need to be treated, and therefore the size and cost of the equipment. If oxygen is used in place of air, however, the combustion temperatures are very much higher, as there is no nitrogen to act as a heat sink. The temperatures are so high, in fact, that this is the process used in an oxy-acetylene torch used to cut steel! In this case, therefore, some of the CO_2 produced in the combustion process is recycled back to the beginning of the process and is fed into the combustion chamber together with the pure oxygen. The CO_2 then acts as an inert "heat sink" during the combustion process, thereby reducing combustion temperatures to manageable levels, the same way that nitrogen does in a conventional combustion process using air as the oxidant. This is the same type of process as the "exhaust gas recirculation" often used to reduce combustion temperatures in vehicle engines in order to reduce the formation of nitrogen oxide compounds. Elimination of nitrogen from the flue gases does greatly reduce the size and cost of equipment required, but this has to be considered against the added cost of providing pure oxygen which is normally obtained from an air separation plant, which also requires a significant energy input. Much further development needs to be done, and demonstration plants need to be built and assessed to determine whether the benefits of oxygen-rich combustion outweigh the costs and energy penalty of providing the oxygen.

In the pre-combustion concept the usual combustion process is replaced by a partial oxidation, or gasification, process. This is the same process used to produce syngas as a precursor to synthesis into liquid fuels, but in this case the goal is to produce a clean-burning gaseous fuel for use in a gas turbine, while at the same time converting all of the carbon in the fuel into a stream of pure CO_2 which can then be diverted for storage. The partial oxidation process in the gasifier first uses oxygen to convert the fuel (usually coal) into the typical syngas mixture of CO and hydrogen. The CO is then reacted with steam in a second vessel using the "water gas shift" reaction to turn the CO into CO_2 and more hydrogen. The CO_2 is then separated from the hydrogen, using one of the techniques just described, and the hydrogen can be burned with air in a gas turbine combustion chamber. The gas turbine exhaust is then primarily a mixture of water vapor and nitrogen, and this hot flue gas is used to generate steam in a boiler for use in a steam turbine.

This combined cycle process is the IGCC process we have already briefly mentioned, and this is used to generate electricity with an efficiency of around 50%, comparable with that of a natural gas-fired combined-cycle plant. The benefit of the IGCC, however, is that it is able to use a low-grade fuel, coal, to generate electricity efficiently and provide a concentrated stream of CO_2 for later storage. Again, however, this process normally requires pure oxygen to feed the gasifer, and the cost and energy consumption of the oxygen plant need to be taken into account when evaluating the overall process economics.

At the present time there is a great deal of research and development being done on all of these processes, and it is too soon to see a "winner." Within the next decade, however, it should be possible to identify one or more of these processes as the most technically viable, and cost-effective, to facilitate CO_2 capture and storage while using coal to generate electricity at high efficiency. In engineering studies, however, the IEA has estimated that the additional costs of adding CO_2 capture and storage to coal-fired powerplants would increase the cost of electricity by between 50% and 100% of the cost without capture and storage, depending on which technology is ultimately used, and on the cost of the fuel. Although these costs are substantial, they do not seem to be out of the question if we are serious about satisfying our growing energy needs using widely available, and low-cost, coal resources without adding greenhouse gases to the atmosphere.

BIBLIOGRAPHY

Alberta Energy and Utilities Board (AEUB) (2005). *http://www.eub.gov.ab.ca/bbs/default.htm*

Canadian Association of Petroleum Producers (2005). *http://www.capp.ca/*

Canadian National Energy Board (2005). *http://www.neb-one.gc.ca/index_e.htm*

International Energy Agency (2001). *Putting Carbon Back in the Ground.*

SASOL (2005). *http://www.sasol.com/sasol_internet/frontend/navigation.jsp?navid=1&pnav=sasol&cnav=sasol*

Suncor (2005). *http://www.suncor.com/start.aspx*

Syncrude (2005). *http://www.syncrude.ca/*

US Geological Survey (USGS) (2005). *http://www.usgs.gov/*

World Energy Council (2005). *http://www.worldenergy.org/wec-geis/default.asp*

7

Renewable energy sources

7.1 INTRODUCTION

Renewable energy sources are primarily those which are inexhaustible in nature, and which are ultimately derived from the radiant energy of the sun reaching the earth. These include the obvious examples of hydroelectric power, solar energy, and wind power, as well as some not quite so obvious examples, such as combustible renewable wastes and biomass fuels like ethanol made from grain crops. In addition, sources such as geothermal energy and ocean gradient energy, which are derived from the very large quantities of thermal energy stored in the earth's crust and oceans, are often categorized as "renewable," although clearly in the very long-term they are not completely sustainable. Of course, if we were to take a time-scale of millions (or perhaps billions) of years, even the sun's radiant energy will diminish, and so none of these sources is truly sustainable "for ever." To a certain extent, then, the definition of "renewable" is somewhat arbitrary, but clearly these are all sources which should still be available to future generations thousands of years from now, and not just in the next few hundred years, as is the case for "non-renewable" sources, such as fossil fuels. Even nuclear power, depending on the technology used to access the energy in nuclear "fuels," is sometimes considered to be renewable, because potentially it will be available for much longer than fossil fuel-derived energy. We will consider nuclear energy to be in a separate category of "sustainable" energy sources, however, and this will be described in the next chapter. Another important consideration for all forms of sustainable energy is the fact that their use will not normally contribute to a net increase in the atmospheric concentration of greenhouse gases such as CO_2.

An important characteristic of most, but not all, sources of renewable energy is the low "energy density," or energy generated per unit cross-sectional area, or surface area. Both solar energy and wind energy, for example, have very low energy density, which means that relatively small quantities of energy are available from each square meter of the earth's surface area. This is notably not true, however, for hydroelectric power, as in this case the radiant energy originating from the sun has enabled the global climate system and the earth's geography to "concentrate" the energy. The power from a hydroelectric installation is then derived from the potential energy stored in a large body of water in an elevated position which is used to drive turbines as it falls to a lower level. The reservoir is continually refilled by rainfall, which is collected over a very large area, and then guided by the natural topography and river systems to run back into the reservoir. In order to use solar energy directly, however, the sun's rays must be intercepted over a very large area, since the incoming "insolation," or incident radiant power of the sun, is only about 1.37 kW/m^2 outside the earth's atmosphere. Due to absorption of some of the energy by the atmosphere this decreases to a maximum value of approximately 1 kW/m^2 at the surface of the earth on a clear day, depending on the time of year and location. This is enough energy to power ten 100 W light bulbs, although since solar photovoltaic panels used to generate electricity directly from sunlight may have an average efficiency of only about 10%, one square meter of panel would be able to power one light bulb on a clear day. During cloudy periods solar energy is still available, but it is considerably less than during clear periods, and of course no solar energy is available at night. The annual average solar energy received on the earth's surface, therefore, is considerably less than indicated by the peak solar insolation value, and varies widely depending on the particular latitude and local climate. For example, in the USA, the National Renewable Energy Laboratory of the DOE (2005) has estimated the average annual solar energy to range from a low of about 4 kWh/m^2 per day in northern and cloudy areas, to a high of nearly 7 kWh/m^2 per day in the sunny southwest. Solar energy can be used directly not only to produce electricity, but also to provide heat for both residential and commercial buildings. Satisfying these heating requirements is made somewhat problematic by the fact that the heating season usually corresponds to the period when solar energy availability is at its lowest. Nevertheless, with the help of some type of energy storage, or with buildings designed to incorporate "passive" solar energy, a significant fraction of the annual heating requirements can be obtained from solar energy.

Similarly, wind energy also has a low energy density, although of course there are some very windy areas which have much higher wind energy potential than others. Again using the example of the continental USA, the average annual wind power ranges from a low of less than 200 W/m^2 in the south-eastern region of the country, to greater than 800 W/m^2 in the Rocky Mountain region. Since wind strength close to the earth's surface increases significantly with height above the ground due to the nature of the planetary boundary layer, these data are standardized at a height of 50 m, and correspond on the low end to an average wind speed of less than 5.6 m/s and range up to a high of over 8.8 m/s on the high end. Wind energy is less evenly distributed than is solar energy, and in the USA the concentrations of high wind energy potential may be quite remote from the major load centers on both coasts and in the mid-west region around the Great Lakes. In subsequent sections of this chapter we will examine in more detail the potential for both solar and wind energy, as well as other less well-developed technologies designed to extract energy indirectly from the sun.

7.2 SOLAR ENERGY

7.2.1 Solar thermal energy systems

One way of utilizing solar energy is to use it directly as a source of thermal energy, either to provide space heating for residential and commercial buildings, or to generate electricity using a conventional Rankine steam cycle. As we have seen, a great deal of energy is used to provide basic comfort in buildings, and in the populous mid-latitude countries this is primarily used for heating during the winter months. The use of both "active" and "passive" solar thermal energy systems for these applications could provide a significant reduction in the need for non-renewable primary energy sources. Passive solar heating simply refers to architectural design techniques which enable the building structure to absorb as much solar energy as possible during daylight hours in the winter months, and then using this "stored" energy to replace heat that would normally be provided by a fossil fuel-fired furnace, or by electric heating. Design concepts can be as simple as ensuring that windows are minimized on north-facing building walls, and enlarged on south-facing walls so that as much sunlight as possible will enter the building and heat up structural elements such as internal walls and floors. More complex design ideas have also been utilized to increase this passive heating, including the use of "Trombe walls," for example. These are heavy, usually black-painted concrete walls placed just behind south-facing glass that are used specifically to absorb as much heat as possible from the sun's rays, so that this thermal energy can be released over periods of several hours. The glass just in front of the wall acts as a greenhouse to trap as much solar energy as possible, and then air is allowed to circulate through the gap between the glass and the concrete. The circulating air then absorbs heat which has been stored in the wall and transfers this to other parts of the room, or even to other parts of the house. The massive wall structure is able to absorb sufficient energy so that heat can be transferred to the circulating air for several hours after the sun has gone down. Some installations have even included blinds just inside the glass which are automatically closed on cold nights in order to reduce the energy which would otherwise be lost by being re-radiated back out through the window.

Active solar heating uses "solar collectors," usually mounted on rooftops for residential buildings, to heat water, or another fluid which is then circulated to other parts of the building. These active solar collectors can also be used as a source of domestic hot water, or to provide heat directly to a swimming pool. The outdoor swimming pool application is particularly attractive, since these are usually used during the warm summer months when the maximum amount of solar radiation is available. The economics of solar water heating are obviously affected by the cost of alternative energy sources used for this purpose, principally electricity and natural gas, and by the building location. In the USA, for example, solar heating of swimming pools is particularly attractive in sunny states like California and Florida in which there are many outdoor swimming pools. In most installations, whether they are used for domestic hot water or for swimming pools, a conventional water-heating system using natural gas or electricity is installed to provide back-up energy during cloudy periods or when cool weather results in extra demand for hot water. In many cases, however, more than half of the cost of traditional sources of energy can be saved over the course of a year using solar energy, and in some cases much more than this. The solar system costs are also reasonably modest, so that financial "payback" times can be less than 10 years, making solar energy an attractive investment.

Finally, the "concentrating solar collector" is an active solar thermal energy installation usually used to generate electricity on a fairly large scale. These systems use one or more reflecting mirrors to concentrate a beam of solar energy onto a focal point in order to provide a source of high temperature heat. The use of a large number of mirrors over a wide area can provide a relatively low-cost source of concentrated energy, suitable for heating water or other fluid to a high temperature. This high temperature heat can then be used either to run a hot-air, or "Stirling" engine, or to provide steam for use in a conventional steam-generating plant, both of which are used to drive an electric generator. Of course, this type of system can only be used to provide a source of thermal energy during daylight hours, although some large systems incorporate a thermal storage system so that they can continue to generate electricity for some time during cloudy periods, or even at night. If a source of firm electricity is required then some form of back-up system may also be required for stand-alone applications. In large-scale demonstration plants built to date in the USA, a "hybrid" system using natural gas as a back-up fuel has been used to provide continuous generation, even during the night. However, one of the advantages of using solar-based systems in hot sunny climates is that the period of maximum electrical output corresponds closely with the period of maximum demand for air-conditioning. Smaller systems using a parabolic mirror with a Stirling engine at the focal point have also been suggested as a possible way to provide electricity for small rural communities in developing countries, and particularly for those in regions with high levels of solar insolation.

Larger installations, using an array of mirrors covering a wide area have usually been funded by government departments or research agencies, and have been built in desert or near-desert conditions to demonstrate the technology. These systems have been built primarily to demonstrate the technology, using two different approaches; either a solar "power tower" concept, or a "trough" concept. A solar power tower thermal plant uses a large number of mirrors, or "heliostats" which are able to automatically track the sun and focus the reflected rays onto a "receiver" on top of the central tower. The receiver is heated to a very high temperature by the highly concentrated solar radiation, and this is used to heat water to produce steam directly, or in some cases to heat molten salt which has a greater capacity to carry this heat away and then transfer it to water in a secondary boiler. In either case the steam which is ultimately produced is then used in a conventional Rankine cycle to drive a steam turbine-powered generator. Figure 7.1 (US Department of Energy, 2005) shows the Solar Two demonstration plant located in the Mojave desert, near the town of Barstow, California. This plant uses molten salt as an intermediate heat transfer fluid, and is a retrofitted version of the original Solar One plant, which

Figure 7.1 "Solar Two" concentrating solar power plant. *Source*: DOE.

heated water directly in the tower to produce steam. The original Solar One plant operated between 1982 and 1988, with a peak output of 10 MWe under clear sunny skies. The molten salt heat transfer fluid used in the Solar Two plant increases the ability to store energy for use during cloudy periods and at night. Successful operation of this plant over a 3-year period from 1996 to 1999 confirmed the benefit of increased energy storage capacity. These results then led to plans for the construction of a similar plant in Spain, the "Solar Tres" (Solar Three) plant, where a substantial renewable energy subsidy makes this an economically attractive option. This first commercial plant is designed to have a peak solar energy input of around 40 MW, and will use molten salt thermal energy storage so that a 15 MW turbine can be operated for 24 hours per day during the summer, with an annual capacity factor approaching 65%.

A newer technology now being demonstrated in the USA, is the solar "trough" concept, which uses an array of parabolic mirrors which focus the sun's rays on a receiver pipe which runs along the complete length of each mirror at the focal point. This concept then does not require a tower, since the heat is collected continuously by the hot oil heat transfer fluid piped around the complete linear mirror array. A heat exchanger is used to transfer heat from the hot oil to boil water with the resulting steam again being used to generate electricity using a conventional Rankine cycle. The parabolic mirrors are all aligned along a North–South axis, and are automatically tilted to follow the

Figure 7.2 Kramer Junction solar trough power station. *Source*: DOE.

sun as it traverses the sky from east to west. The mirrors are therefore focused on the sun for the maximum possible time to maximize the amount of solar energy collected. Nine of these solar electricity generating systems (referred to as "SEGS") have been built in the Mojave desert, ranging in size from 14 to 80 MWe peak power. In total these provide a peak electrical output of some 350 MWe, with the power being fed into the California grid. Natural gas is used in a "hybrid" fashion so that firm electricity can be generated during extended cloudy periods, or at night, but to date abut 75% of the total electrical energy produced has been generated from solar energy. These plants have a lower capital cost than the solar power tower design, and have shown that they can provide the cheapest form of solar-generated electricity, at around $0.12 per kWh. A partial view of the Kramer Junction solar trough power station operating in California, which consists of five individual plants (SEGS 3 to 7) generating a total of 150 MWe, is shown in Figure 7.2 (US DOE, 2005). This gives an indication of the scale of the individual solar concentrating troughs and shows the receiver pipe running along the focal point of each trough.

7.2.2 Photovoltaic solar electricity generation

Photovoltaic (or PV) solar cells are manufactured from special semiconductor materials that use the energy of the photons from solar radiation striking the cell to produce an electric current. The "photovoltaic effect" results in electrons being separated from individual atoms when these photons strike the cell material, and the flow of these "free" electrons through the material will generate a voltage of approximately 0.5 volts. This voltage can then generate an electric current which is supplied to an external load. The most common material used for manufacturing PV cells is silicon, which is usually doped with phosphorus or a similar material to ensure that free electrons are released when the material absorbs the incident photons. The cells also incorporate a conducting metallic mesh so that as many of the free electrons as possible can be collected, and then routed through the external load. The most expensive solar cells are made from crystalline silicon wafers which are cut from a single crystal which has been specially "grown." These have the highest efficiency of conversion from solar radiation to electricity of any solar cells, although this value is still only about 15%. One reason for the inherently low conversion efficiency is that most of the sun's energy is contained in the long wavelength part of the solar spectrum which does not result in photons being absorbed by the cell. Polycrystalline silicon, which is easier to manufacture and therefore lower in cost, is also used, although the energy conversion efficiency of the resulting solar cells is less than that of single crystalline cells. So-called "amorphous silicon" solar cells are manufactured using thin-film techniques, but their efficiency is only about one-half that of crystalline cells. There is an obvious trade-off between cost of production and efficiency of conversion of solar radiation into electricity, and this is an area which is under very active development.

A typical PV solar "panel" consists of many individual solar cells connected together so that enough current can be generated to provide power to the external load. The efficiency of these panels, defined as the electrical power output divided by the solar insolation input, is around 10–15% for most commercial crystalline silicon PV panels, and about one-half of these values for the cheaper amorphous silicon panels. Groups of individual cells are connected together in series to increase the voltage, usually to between 12 or 24 V DC, and then these groups are connected in parallel to form a complete panel. A typical solar panel measures about 1.5 m × 0.8 m, which is easy to handle, and if the cells are made from crystalline silicon this will usually have an output in bright sunlight of around 150 Wp (where Wp indicates peak Watts). A typical US home uses 5000 kWh of electrical energy per year, or on average nearly 15 kWh per day. In a region with an average insolation of 5 kWh/m^2 per day, this indicates that a solar collector area of some 30 m^2 would be required to meet the total electrical requirements, assuming an average efficiency of 10%. The 30 m^2 of PV panel might have a peak electrical output of around 3.5 kW, which should accommodate most of the electrical load from the house while it is in direct sunlight. However, there is usually a mismatch between peak generating capacity and household electrical demand. For example, the generation of electrical power will peak around mid-day on a clear summer day when all of the residents may well be at work or on the beach. And, of course, peak electrical demand may occur around nightfall on a dark mid-winter day when there is little or no availability of solar energy. In each case there needs to be some type of battery storage system available, so that the energy generated during the summer peak is not wasted, or to ensure an adequate supply of electricity at night and in the winter months. Or, there needs to be access to a "back-up" electrical system to ensure adequate electricity supply at night and during cloudy periods in which there is a heavy demand, or to accept excess energy which may be generated during the summer when demand is low.

The requirement for either storage or back-up from an electrical grid adds an additional complication, and usually significant cost, to the solar PV electricity system. Also, the intermittent nature of the solar energy means that the PV system is only able to generate peak levels of electricity for a relatively short period during any one year. This results in a low "capacity factor," perhaps better described as a "utilization factor," which is defined as the ratio of the annual energy generated to the amount which would be generated if the system were to generate at peak output, 24 hours per day for a complete year. Using this definition for a conventional fossil fuel-fired, or nuclear power-plant for example, a typical capacity factor might be 80–90%. In other words, the plant is expected to operate at 100% peak output for 80–90% of the time, with the remaining time taken up by maintenance or "forced outages." For a solar PV system in mid-latitudes, however, this factor may be as low as 10%, or even lower in some areas, due to the limited availability of the primary solar energy source over the year. There will be no solar energy available at night, and less than 100% output during the day when the PV system is partially obscured by cloud. In the UK, for example, which has quite a cloudy climate for much of the year, the UK Energy Saving Trust has estimated that a 1 kWp PV system should generate a minimum of 750 kWh per year. This corresponds, however, to a capacity factor of only 8.5%, making it less than an ideal location for solar PV production. The implication of such a low capacity factor is that the capital equipment is poorly utilized, so that the capital costs per unit output of electricity are greatly increased. This factor, together with the high initial cost of PV panels, significantly increases the unit cost of solar PV generated electricity in comparison to the costs of conventional generation in most parts of the world.

It is not easy to obtain detailed cost and performance information for a typical installation, but the Lord house, located in the eastern US state of Maine (Maine Solar House, 2005), and used as a case study by the IEA, can be used as a good guide to the current economics of solar PV electricity generation. This house was designed with energy efficiency and sustainable energy as a focal point, and so it is well-insulated with fairly small windows and is positioned so that the solar panels on the roof face to the south with no shading. The roof contains both solar thermal collectors for space heating, and 384 square feet (35.7 m^2) of solar PV panels. All of the data provided here was obtained from the IEA PV electricity website (International Energy Agency, 2005) or the website maintained by the owner, Mr. William Lord (Maine Solar House, 2005). One factor which made this installation quite attractive was the net metering policy adopted by the local utility to account for the backup provided by the grid-connection to the house which was maintained. This policy states that only the net electricity taken by the house would be charged, while any net electricity provided to the grid would not be credited. In this case, since over the complete year the PV system provided a net 591 kWh to the grid, there was no charge for electricity provided by the power company, while they received 591 kWh for free. There was, however, a small monthly connection charge of $8.00 per month to help defray the utilities costs.

The main data relating to the house for the calendar year 1998 as provided by the owner are shown at the top of Table 7.1, and these were used to generate the "derived data" shown at the bottom of the table. From the bottom half of the table it can be seen that the total "avoided cost" of electricity that would have been purchased in the absence of the PV system was about $439 based on a consumption of 3655 kWh for the year and a unit cost of $0.12 per kWh. The total system cost of $30 000 was estimated by the author using a PV panel cost of $5.00 per peak watt which is the approximate unit cost today, and a rough estimate of $1000/kW for the inverter. The balance of $4800 was assumed for metering and control equipment, as well as system installation. The very simple message from Table 7.1 is the fact that had the $30 000 total system cost been invested at an interest

Table 7.1. *Lord solar house data – 1998*

Input data	
Peak electrical power	4.2 kW
PV area	35.7 m^2
PV electricity generated	4246 kWh
Electricity supplied to the grid	3008 kWh
Electricity taken from the grid	2417 kWh
Net electricity provided to the grid	591 kWh
Unit cost of grid electricity	$0.12 kWh
Derived data	
Annual electricity consumption	3655 kWh
Capacity factor	11.5%
Unit cost of PV panels (estimated)	$5.00 Wp
Total cost of PV panels (estimated)	$21 000
Cost of inverter (estimated)	$4200
Installation and controls costs (estimated)	$4800
Total system cost (estimated)	$30 000
Avoided cost of electricity	$439
Interest rate	5%
Return on system cost if invested	$1500

rate of 5%, the return on investment would have been more than three times the avoided cost of electricity provided by the system. Even with an interest rate of 3% the investment income would be more than double the avoided electricity cost. Within the current economic conditions, therefore, and with system cost assumptions and current residential electricity rates shown in Table 7.1, the solar PV system is clearly not economically attractive at the present time. However, this doesn't take into account the benefit of generating electricity free of any greenhouse gas emissions, or any changes in the future which may accrue due to decreased cost of PV systems and any increases which may occur in the cost of residential electricity.

The system just described using solar PV generated electricity to supply the needs of an individual residence is one example of a "distributed energy" system, in which the electricity demands from many small-scale users are met, at least in part, on-site rather than being supplied from a large utility system. These systems can be attractive, particularly when the primary energy source, in this case solar energy, is very diffuse in nature with a low energy density. The widespread adoption of distributed energy systems shifts the burden of supplying

Figure 7.3 Springerville solar PV generating station, USA. *Source*: Tucson Electric Power.

the capital funds needed for generation equipment from large utility companies to individuals, or small businesses, but then relieves them from making regular energy payments. As such, the type of financing for these systems can be a critical factor in determining the rate of adoption. As we have seen from the example of the Lord house, for most individuals the financing of a residential solar PV installation would not be an economically attractive proposition given today's costs and electricity rates in most industrialized countries. Most private installations in the developed world today are funded by "early adopters" who are interested in demonstrating the environmental advantages of renewable energy, rather than in saving money. In order for this to change there will have to be a significant decrease in the costs of PV panels and inverter costs, and/or a large increase in utility costs due to requirements to eliminate, or at least reduce, greenhouse gas emissions.

Primary energy sources with low energy density can also be used for a more centralized generating system, however, if that is seen to be a more appropriate way to finance large-scale renewable energy development. The largest centralized solar PV powerplant in the world, shown in Figure 7.3, is the Springerville Generating Station (Tucson Electric Power, 2005) in the USA. This plant is located in the Arizona desert, one of the sunniest locations in the continental USA, and has a peak generating capacity of 4.6 MWe. The plant incorporates nearly

35 000 solar panel modules with a total coverage area of 44 acres, or 17.8 hectares, with the panels fixed at a tilt angle of 34 degrees and facing due south. Although there are only limited operating data available for the station at this time, the total annual electrical energy production in 2004 was 7 064 000 kWh, which corresponds to a capacity factor, or utilization factor, of 17.5%. This is significantly better than the 11.5% achieved by the Lord house in Maine, or the 8.5% minimum estimated for the UK, and illustrates the benefit of locating such a plant in a sunny area at lower latitudes. When evaluating various ways of generating electricity with zero greenhouse gas emissions, utilities will compare the cost of using large-scale solar PV installations with the cost of generation from sources such as nuclear power, or perhaps fossil fuel generation with CO_2 capture and storage. Two important considerations will be the capital cost of the generating equipment, in dollars per installed (or "peak") kilowatt, and the capacity factor. Taken together, these two factors will largely determine the annualized cost, in $/kWh, of the electricity generated. At this time the capital cost of solar PV generation, at some $5000 per peak kW is about 2.5 times that of a nuclear powerplant, which is about $2000 per peak kW (see Chapter 8). The capacity factor, however, is expected to be about 85% for the nuclear plant, while as we have seen above it is only about 18% for the Springerville solar PV station. Of course the cost of fueling the nuclear plant would have to be taken into account in any proper economic comparison, while the primary energy cost for the solar PV plant is free, but this is a relatively small cost for a nuclear plant. The combination of higher capital cost per peak output capacity, and the very low capacity factor, or utilization factor, for the solar plant makes the annualized cost of electrical energy generated much higher than for the nuclear plant. This low capacity or utilization factor, together with the need for back-up power or large-scale energy storage, will be one of the major challenges associated with all forms of intermittent energy systems, including both distributed and centralized solar PV systems, as well as wind energy systems.

A major challenge to widespread adoption of intermittent renewable energy sources, and particularly solar PV which produces no electricity at night, is the need to have access to large energy storage capacity, or to a large electrical grid capable of providing back-up. In the long term, if PV systems become more economically attractive, and therefore much more widespread, there may be additional pressures placed on utilities asked to provide access to the grid as a back-up device. This will be true for both distributed energy systems, as well as for larger centralized plants, such as the demonstration plant in Tucson, Arizona. The small fixed monthly connection charge used in the Maine residential case study discussed previously probably does not capture the true costs of providing back-up power whenever required. When the demands to provide significant levels of back-up power increase, accompanied by very little or no revenue from electricity sales, utilities may have to introduce higher connection charges, which could then adversely affect the economics of PV solar systems. There is likely to be a limit therefore, on the total amount of intermittent renewable energy generation which can be "absorbed" economically by a utility system. This limiting amount of generation capacity will vary from utility to utility, depending on the particular demand profile of the utility and whether or not it has significant levels of energy storage, such as that provided by a large share of hydroelectric power capacity.

7.3 WIND ENERGY

We have noted that wind energy also has a relatively low energy density, and its potential is quite unevenly distributed. Wind energy has been used for several centuries, initially in the form of windmills used to provide power for milling grain and to drain low-lying land in the Netherlands and parts of England. In the early part of the twentieth century before rural electrification made utility-generated electricity widely available, many farms in North America used small-scale windmills to generate electricity locally. These all but disappeared, however, as cheap electricity from large-scale powerplants became widely available around the middle of the century. As we enter the twenty-first century wind power has made a comeback and is currently the most significant source of renewably generated electricity (other than hydro power). The new windmills (or wind turbines as the manufacturers now prefer to call them) are much larger than in the past, and are now available in unit sizes up to 4.5 MWe peak capacity, with units up to 5 MWe now in the development stage. In Europe, in particular, there has been widespread adoption of wind-power as a source of electricity for major utilities, with Germany and Denmark leading the way in the use of wind energy. Germany has the largest installed wind energy capacity in the world, while Denmark produces nearly 20% of its total electrical energy from wind-power. There has been tremendous growth in wind energy capacity in recent years, particularly in Europe, as can be seen in Table 7.2 (European Wind Energy Association, 2005).

Table 7.2. *World wind energy capacity (MWe)*

	2001	2003
Germany	8734	14 612
USA	4245	6361
Spain	3550	6420
Denmark	2456	3076
India	1456	2125
Italy	700	922
UK	525	759
Netherlands	523	938
China	406	571
Japan	357	761
Rest of the world	1975	3756
Total	24 927	40 301

Source: European Wind Energy Association.

However, due to the low capacity factor for wind power, the amount of energy contributed from this source is still only a very small fraction of the total demand for electricity. In 2003, for example, the total amount of energy generated from wind was only about 0.5% of total worldwide electrical energy production.

The large wind turbines currently being installed make use of modern lightweight materials to reduce their weight and cost, and have also benefited from modern generator and control system design to improve their performance. These turbines usually make use of variable pitch controls for the rotor, as well as variable speed gearless generator designs to accommodate varying wind speeds. A photograph of one of the largest wind turbines currently available, an Enercon E112 model being installed near Magdeburg, Germany, with a maximum capacity of 4.5 MWe, is shown in Figure 7.4 (Enercon, 2005). This very large unit has a rotor diameter of 114 m, providing a swept area of 10 207 m^2 with a hub height of 124 m above ground level. Having such a large hub height provides a significant advantage, in that the rotor is placed higher in the planetary "boundary layer" where the average wind speeds are much greater. The rotor is constructed of lightweight glass reinforced epoxy resin, and turns at a variable speed between 8 and 13 rpm. The turbine is designed to operate with a minimum wind speed of 2.5 m/s, and has a "cut-out" speed, to avoid damage to the turbine and generator, of some 30 m/s.

Figure 7.4 Large 4.5 MW wind turbine. *Source*: Enercon.

Although many wind turbines have been installed as "one-off" installations, primarily to demonstrate the technology, the trend now is to build "wind farms" with many wind turbines situated in an area with high average wind speed. These wind farms may be located on land, usually in remote areas where there is little interference with human activity, but increasingly "off-shore" wind farms are being built in shallow sea-bed locations in coastal areas. There are additional construction challenges with the off-shore locations, of course, as underwater foundations and the towers must be built to withstand severe wave action, as well as high wind speeds. There are significant benefits, however, in that wind speeds are usually much higher in coastal areas where the open water enables the wind to build up with little interference. Off-shore design and construction techniques have also benefited from the large experience gained over many decades in building off-shore oil and gas extraction facilities. Another benefit of off-shore locations is that the turbines are usually located well away from significant human activity, and therefore tend to be more acceptable to the local population. A photograph of one of the largest off-shore wind farm installations to be built so far, at Middelgrunden near Copenhagen in Denmark, is shown in Figure 7.5 (Middelgrunden Wind Farm, 2005). This impressive installation, in the Oresund strait, which separates Denmark from Sweden, consists of 20 turbines providing a total capacity of 40 MWe, and provides nearly 4% of the annual electrical energy consumption of Copenhagen.

In considering the contribution from wind energy, like that from solar energy or any other intermittent energy source, one needs to be careful not to confuse the power *capacity* of the wind turbine, with the amount of *energy* which is generated. The capacity quoted for a wind turbine represents the maximum amount of power (in MWe) which can be generated by the turbine at the design wind speed. Due to the intermittent nature of wind energy, however, wind speeds at, or in excess of, the design wind speed occur for only a fraction of the year. The intermittent nature of the contribution of wind-power to electrical energy generation is accounted for by the "capacity factor" of the turbine. The capacity factor represents the fraction of the energy actually generated (in MWh) by the wind turbine over a particular period (usually one year) to the maximum energy which theoretically could be generated if the wind blew at or above the design wind speed for the whole year. Usually this factor is between 20% and 30% for individual turbines, although it naturally depends on the particular site. For the year 2003 the IEA has estimated (IEA, 2005) that total worldwide wind energy production amounted to 84.7 TWh, or 0.51% of the 16 666 TWh of electricity generated from all sources. Using the total installed wind capacity of 40 301 MWe in 2003 shown in Table 7.2, the overall worldwide capacity factor for wind turbines in 2003 was 24%. In comparison, a large fossil fuel-fired powerplant, or nuclear power station, will normally have a capacity factor of between 80% and 90%, indicating a

Figure 7.5 Off-shore wind farm at Middelgrunden, Denmark. *Source*: Middelgrunden Wind Farm. Photocopyright, Adam Schmedes, Lokefilm.

much better utilization of the capital equipment on an annual basis. The low capacity factor for wind turbines also shows the need to carefully select a new wind-farm site to ensure that it is in an area with a large fraction of high-wind days.

The intermittent nature of wind power also necessitates that substantial reserves of "back-up" power, or energy storage is available to ensure reliable electricity supplies during periods of low wind activity. This will usually not be a major issue when the wind power capacity is a small fraction of total system capacity, as there is usually sufficient spare capacity to ensure that the total power demand can be met. In order to replace the firm capacity of fossil-fuel or nuclear plants with wind turbines, the installed capacity needs to be much greater than

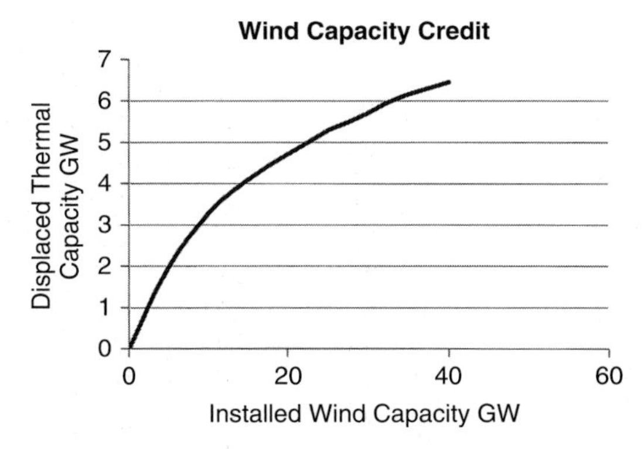

Figure 7.6 Baseload capacity displacement with increasing wind penetration. *Source*: Grubb, M. J. (1986). The integration and analysis of intermittent sources on electricity supply systems. Ph.D. thesis, Cambridge University.

that of the plants they are replacing. One of the issues, of course, with any intermittent source of power is the concern that there may be no power available during periods of high demand, due to a lack of wind, for example. Proponents of wind power argue that as long as the installed wind-power capacity is geographically diverse there should always be at least some contribution available from wind turbines. Studies have shown that the "capacity credit" for wind power used to replace baseload thermal powerplant capacity should be proportional to the square root of the installed wind capacity, as shown in Figure 7.6 (Grubb, 1986). For example, it would require approximately 9 GWe of wind capacity to replace 3 GWe of nuclear or coal-fired power capacity. Although Denmark has achieved a penetration of 20% wind energy into their electrical grid this would not be possible without electrical ties to the neighboring countries of Norway, Sweden, and Germany which can be used to provide back-up power when wind speeds are low.

The recent rapid expansion in wind power worldwide is due in part to the technical advances in design and construction of large multi-megawatt turbines, which has led to a lowering of the unit cost of wind-generated electricity. This cost reduction is enhanced when these large turbines are grouped together in wind farms, which have additional economy-of-scale effects on reducing electricity costs. The expansion of wind-power capacity has also come about as a result of the increasing costs of fossil fuels, particularly natural gas, which have traditionally been used for electrical power generation, as well as the environmental effects associated with burning these fuels. The current capital cost of building a large-scale wind farm is approximately \$1000–2000 per installed MWe, which is comparable to that of a coal-fired powerplant. These costs cannot be used directly to compare the cost of energy production, however, as the much lower capacity factor associated with wind turbines means that the capital costs have a much greater effect on the final unit price of electricity. The lower capital cost per kWh generated for a fossil fuel powerplant is only one component of the unit electricity cost, with the cost of fuel a major factor. The current unit cost of wind-generated electricity has been estimated to range from US 5 cents to 12 cents per kWh, but is very much dependent on the particular site chosen. Wind-power has also been encouraged in many regions by either direct or indirect government subsidies. In the USA, for example, a federal tax credit of 1.5 cents per kWh may be obtained for the first 10 years of wind-generated electricity production. At the lower end of the cost estimates wind-power is now competitive with most fossil fuel-generated power, and this is particularly true with the recent rapid increases in the price of natural gas. These costs will likely be further reduced over the next decade, and wind-power will become an important component of the generation mix in many electrical utilities.

7.4 BIOMASS ENERGY

Biomass energy was the very first form of energy used by humans, and the burning of wood gathered by hand is still an important source of heat for cooking and space heating in many underdeveloped parts of the world. Even in more industrialized countries, particularly in rural regions, wood-burning fireplaces and stoves are often used to provide at least some component of a family's space heating requirements in the winter. The use of biomass energy has now grown much beyond its humble beginnings as a domestic fuel, however, and is used in many different forms in a wide range of industries. These include, for example, combustion of wood-waste to generate steam in pulp and paper mills, the use of "landfill gas" from municipal solid waste (MSW) for electrical power generation, and the production of "biodiesel" fuel and ethanol from corn and grain crops. The direct combustion of wood, and other biomass fuels, such as MSW and agricultural wastes, still accounts for by far the largest component of current biomass energy use. We have already seen in Figure 5.1 that biomass fuel, in the form of "combustible renewable wastes," or CRW, made up just over 10% of total world energy production in 2002. In developing countries, however, biomass-derived energy usually makes up a much greater fraction of overall energy use, and can even be the dominant energy source in some of the poorest countries.

Many studies have suggested, however, that biomass-based energy will provide an even greater share of the overall energy supply as the price of conventional fossil fuels increases over the next several decades. Proponents of biomass energy also point out that the use of biomass as a source of energy is very attractive, since it can be a "zero net CO_2" energy source, and therefore does not contribute to increased greenhouse gas production. The combustion of biomass energy does result in the production of CO_2, however, since nearly all of the carbon in the fuel is converted to CO_2, just as it is during the consumption of fossil fuels. The zero net CO_2 argument relies on the assumption that new trees, or other crops, will be replanted to the extent that they will absorb any CO_2 released during the consumption of biomass energy. This may well be true for properly managed "energy plantations," but is not likely to pertain in many developing countries where most of the biomass energy is obtained from forests which are not being replanted, at least not to the same degree that they are being harvested. Also, the widespread expansion of biomass energy use may result in significant concerns about the availability of land, which may otherwise be used for food production, or other commercial uses such as timber production. One recent review of 17 biomass energy studies showed a wide range of estimates of future biomass energy potential, ranging from the current level of approximately 42 EJ (or 1 Gtoe), to nearly 350 EJ, close to the current level of total energy production, by the year 2100 (Berndes *et al.*, 2003). The wide range of estimates is due, in large part, to the very different assumptions made for both land availability and crop yields.

Combustion of wood-waste, including sawdust, bark, and other residue, is a well-established technology and widely used to generate heat and electricity in the wood-processing industries. This is often done in a "cogeneration" plant in pulp and paper mills in which steam is first used to generate electricity using a steam turbine, and the exhaust steam is then used to provide heat for the process. Also, "black liquor" from kraft pulp mills, which consists of lignin removed from the wood chips during the pulping process together with spent chemicals, is often burned to generate both electricity and process heat. The special "recovery boilers" are so-called because they are also used to recover some of the chemicals contained in the black liquor for re-use in the process. Some researchers have suggested that this use of wood for production of heat and electricity could be greatly expanded by utilizing solid wood from fast-growing tree plantations, rather than relying on waste material from forest products operations. This could be made to be a sustainable operation, with little or no net production of CO_2, if as much forest is replanted as is used to provide energy, as we have discussed above. Combustion of Municipal Solid Waste (MSW) is also now widely used, both as an effective way to dispose of domestic refuse, and as an important source of heat and electrical power. In some cases this is done by burning the MSW in specially modified steam boilers that are able to handle the fuel composition variability and high moisture content of MSW, and process the large quantities of ash that are formed. These have been particularly successful in Europe, where the relatively high price of conventional energy, and the high population density, has provided additional incentives to process domestic refuse in this way. In addition to the use of conventional steam boilers, there is increasing interest in the use of pyrolysis, or gasification technology to produce a combustible gas from MSW. This would often be of interest to smaller communities, or small industrial operations, where the volume of MSW, or other biomass waste such as chicken litter, is not sufficient to justify the cost of a large steam plant.

Another approach to the use of MSW as an energy source is to capture the methane gas that is produced as a result of decomposition of the biomass material contained in landfills, which are used to dispose of most domestic refuse. This gas can be used to provide a source of heat for nearby greenhouses, for example, or can be used to fuel an internal combustion engine, or gas turbine, which is then used to generate electricity. The production of methane gas occurs because of anaerobic digestion, or decomposition of the biomass material in the absence of air. This occurs naturally at large landfill sites, and the methane can be an important source of greenhouse gas emissions if not captured and used as a source of energy. On a smaller scale, use can also be made of purpose-built anaerobic digesters, which process a steady stream of biomass waste material such as animal manure. These have been successfully used on some farms, for example, to deal effectively with large waste streams that would otherwise be damaging to the environment, or difficult to contain. The resulting fuel, or "biogas," can be used to provide heat in colder climates, or as an engine fuel to generate electricity.

One of the attractions of biomass energy is the possibility that biomass-derived liquid fuels may be used to substitute for gasoline and diesel fuel in transportation applications. This is now a very limited market, but small quantities of ethanol are being used to blend with gasoline, and vegetable-derived oils are being used to substitute for diesel fuels on a small scale. Vegetable oil, sometimes in the form of waste oil from deep-fat fryers, is usually blended with diesel fuel, but can also be used on its own. Ethanol is produced by fermentation of corn or other grain crops, just as it is for the production of alcoholic beverages. Some studies have shown, however, that the production of ethanol is itself an energy-intensive process. Life-cycle assessment analysis has shown that large quantities of energy are required for the distillation process to separate the alcohol from water, and also for corn production in the form of tractor fuel and fertilizer production. Pimentel and Patzek (2005) found that the production of ethanol required between 29% and 57% more fossil energy than is produced in the form of ethanol, depending on the biomass source chosen. This is clearly not a sustainable process, and means that economic large-scale production of fuel ethanol by fermentation may be in doubt. Other studies, however, have indicated that the use of waste cellulosic feedstock such as corn stover for the production of ethanol may be more energy efficient since the lignin in the feedstock can be used as an energy source during the ethanol production process (see Sheehan *et al.*, 2004). Biodiesel fuel, in the form of vegetable oil, may be obtained from sunflowers, soybeans, or rape seed. Although some early studies indicated that the production of biodiesel might also consume more fossil fuel energy than that contained in the resulting fuel, these conclusions have been refuted by more recent studies (see Sheehan *et al.*). It appears, however, that much more work is needed before we clearly understand which liquid fuels derived from biomass, and which production processes, may result in more sustainable substitutes for liquid fossil fuels in the long term.

7.5 HYDROELECTRIC POWER

Hydroelectric power generation is one of the largest uses of renewable energy to date, and is beneficial because the production of "hydro" power produces no greenhouse gases, or other air emissions. The generation of electricity from large-scale hydroelectric powerplants is a well-established mature technology, and is used by utilities around the world as an economic source of renewable energy. Hydroelectric power generation relies on the flow of large quantities of water through hydraulic turbines, which can be up to 700 MWe in size. These installations can be "high head" developments which rely on water falling from a considerable height through turbines located downstream of a large storage reservoir, or may be "low head" or "run-of-the-river" designs in which power is generated by the flow of very large volumes of water through turbines immersed in a river. Hydroelectric plants, particularly the high-head variety, can take up large areas of land for the storage of water behind dams, and are often located in quite remote areas at some distance from major population centers. The development of such facilities is necessarily dependent on local geography, and most major hydroelectric facilities are located in countries with mountainous terrain and many lakes and rivers. Hydroelectric power now accounts for nearly 18% of all electricity generation worldwide, while Canada, the world's largest producer of hydroelectric power, provides nearly two-thirds of the country's total electricity requirements from hydro installations. China, too, is relying on hydroelectric power as one of the major sources to supply the rapidly increasing demand for electricity. The 26 turbines that will eventually be installed in the Three Gorges hydroelectric development on the Yangtze river, for example, will have a peak capacity of 18 GWe when finally completed in 2009 after a 17-year construction period. Although the development of this massive project, the world's largest hydroelectric project to date, has been controversial, it will be a major source of renewable electricity to supply China's fast-growing economy.

Although the capital costs of hydroelectric powerplants are usually higher than those for thermal power stations, hydroelectric plants normally have a much longer life expectancy, and with no fuel costs, provide a low-cost source of electricity. The development of new large-scale hydroelectric plants near to large population centers, and therefore regions of high energy demand, is now somewhat limited, however, as in many parts of the world most of the cost-effective hydro power has already been developed. This is true of the USA, for example, which has one of the largest hydroelectric capacities in the world, but now has very few large potential sites still undeveloped. Attention has been turning in recent years, therefore, to small-scale hydroelectric installations, which are often community based in rural or fairly remote areas. These installations are typically less than 1 MWe in capacity, and do not usually involve construction of a dam, but rather rely on the flow of water in small rivers or streams. These small-scale, or "microhydro" (less than 100 kWe) power plants are often built as stand-alone units to supply a small community, or perhaps a farm or small business, without connection to a utility grid. Such small-scale hydro installations can be very environmentally benign, since they have the same benefit of zero greenhouse gas production found in large-scale installations, but usually none of the environmental and social concerns associated with large-scale projects where there may be widespread flooding of river valleys and displacement of some of the local population. Just as for large-scale hydro power, however, the opportunities for new small-scale hydroelectric power development are very site-specific. There is considerable activity now under way, both by governments and non-profit organizations, to try to identify sites where small-scale hydro power may be a cost-effective alternative to more conventional sources of electricity.

7.6 OCEAN ENERGY

Although they operate on quite different fundamental principles, tidal power and wave power are often considered together, and perhaps both should be referred to as a branch of "ocean energy." Tidal power is somewhat unusual, in that it is a renewable energy source that does not rely on the energy of the sun for its fundamental driving force. In the case of tidal power, it is the variation in gravitational pull resulting from the moon orbiting the earth that provides the driving force. Another attractive characteristic not shared by other renewable energy sources is that it is completely predictable in nature, since the movement of ocean tides is readily predicted from the relative movement of the earth and the moon. There are two ways in which this predictable variation in ocean elevation can be harnessed as a renewable energy source. The first way to harness tidal power is to use some form of dam, or "tidal barrage," to trap large quantities of water that flow into a tidal basin. As the tide then ebbs, the elevation difference between the flooded basin and the outgoing sea level can be used to drive water through a low-head hydraulic turbine, similar to those installed in a large hydroelectric plant. In some cases, the turbines can also be arranged so that power is generated during the flooding of the basin behind the barrage, as well as during the ebb flow as the basin empties. Again, this type of installation is very much affected by the local topography, with natural river estuaries in regions of high tidal variations providing the most cost-effective sites for tidal power. The only tidal barrage powerplant of any significant size to have been built in this manner is the LaRance station near St. Malo on the coast of France. This plant, which entered service in 1966, takes advantage of a nearly 8 m tidal range, and has a peak generating capacity of 240 MWe. Reversible blade "bulb turbines" are used, so that some power can be generated during flood-tides as the reservoir is filled, while most of the power is generated during emptying of the reservoir as the tide ebbs. The turbines generate approximately 610 GWh of electricity per year, which results in a capacity factor of just less than 30%. A much smaller demonstration plant at Annapolis Royal in the Bay of Fundy region in Canada has a capacity of 20 MWe, and with generation of some 50 GWh of electricity per year also has a capacity factor of around 30%.

Other large-scale tidal barrage powerplants have been studied, including proposals for very large plants in the Severn estuary in southwest England and in the Bay of Fundy in Canada, but so far none have been constructed. The main reason for this is the high capital cost, leading to high electricity costs, particularly in comparison to a hydroelectric plant of similar scale and cost. This high cost relative to a hydroelectric plant may be explained by two main differences between a tidal barrage plant and a conventional hydro plant. The first is the fact that the head for a tidal plant is necessarily limited to the local tidal range, which is usually much less than that for a typical hydroelectric plant, and this severely restricts power output. The second difference is due to the intermittent nature of the tidal action, which means that the tidal plant is only capable of generating at peak capacity for a relatively short period of time when the water trapped behind the tidal barrage has reached its maximum level. As the trapped water flows back into the sea the available head is continually reduced, resulting in a lower generating capacity. For this reason the capacity factor for a tidal powerplant (as we have noted for both the LaRance and Annapolis plants) is limited to approximately 30%, which is much lower than for most large hydroelectric plants. This low capacity factor results in poor utilization of the large capital investment normally required for the civil works needed to trap sufficient water behind the barrage, and drives the cost of producing electricity even higher.

The second way to capture energy from the power of the tides is to make use of the energy contained in tidal currents which are regularly formed in narrow coastal restrictions as a result of the periodic change in ocean levels. These currents are also very site-specific in nature but can contain significant amounts of energy on a regular daily basis. Power can then be extracted by immersing one or more turbines into the tidal stream. Turbines suitable for this application are usually similar in design to wind turbines, and both horizontal axis and vertical axis designs have been tested. Tidal current energy technology is still at a very early stage of development, and only a few demonstration projects

Figure 7.7 Marine current turbines "Seaflow" 300 kWe tidal current generator. *Source*: Marine Current Turbines.

have been built. Probably the largest demonstration of tidal power has been the "Seaflow" experimental turbine developed by Marine Current Turbines, and shown in Figure 7.7 (Marine Current Turbines, 2005). This horizontal axis machine has operated some 3 km off the Devon coast near Lynmouth, England since early 2003. It utilizes a two-bladed rotor 11 m in diameter, and is capable of generating 300 kWe. The turbine is attached to a single large vertical pile which is driven into the sea bed, and can be raised above the water surface for inspection and maintenance, as can be seen in Figure 7.7. This installation will be followed by a twin-rotor design, which will be designed to operate with the current flowing in both directions as the tide changes from flood to ebb conditions, and will have a peak capacity of 1 MWe. Other tidal current demonstration projects are being planned, including some that utilize a series of vertical-axis turbines located in a "fence" structure that would help to increase the flow velocity through the turbines in regions of high tidal currents. One of the concerns, of course, with locating marine current turbines in areas where other marine activity takes place is the possible hazard to shipping. The design and installation of tidal current turbines on any kind of large-scale basis will therefore have to be considered carefully on a site-by-site basis.

There is also a great deal of energy contained worldwide in ocean waves, and a wide variety of machines for extracting some of this energy in a practical way has been proposed. None of the technologies considered so far has reached the commercial scale of operation, but research and development continues in many maritime nations. Wave power in the open ocean has been estimated to be as high as 90 kW per meter in the North Atlantic ocean, but of course this varies considerably with location and time of year. The UK, a nation with a long and proud maritime history, has probably been the leader in this area, with much of the activity being concentrated in Scotland and northern England. Wave extraction devices may be broadly classified into on-shore developments, aimed at extracting energy as waves impact the shoreline, and off-shore devices which rely on wave action further out to sea. Although on-shore devices are attractive due to their simplicity, they suffer from significant reduction in power generation potential due to the attenuation of energy as the waves reach the shore. One of the best developed on-shore technologies is the use of wave-action to compress air in a partially enclosed cavity. Most of these devices, known as "Oscillating Water Column" (OWC) devices, first focus the incoming wave energy in order to generate an oscillating column of water in a shore-based facility. The oscillating water column then compresses air in a duct and this is used to drive an air turbine for the generation of electricity. Because of the continuously reversing airflow, a "rectifier" is sometimes used to convert this into a continuous flow pattern to drive a simple turbine. Another approach is to use a special type of turbine with symmetrical blades, so that airflow in either direction will still drive the turbine in the same rotational direction.

The best-known on-shore wave energy conversion device is probably the Limpet (Land Installed Marine Powered Energy Transformer) project pioneered by Queen's University, Belfast and the UK company Wavegen. Following the design and construction of a 75 kWe prototype plant, a much larger demonstration project, designed to generate 500 kWe at peak output, was built on the island of Islay off the west coast of Scotland. This larger Limpet installation was completed in the year 2000, and a schematic of the installation is shown in Figure 7.8. Initial results from the plant have been disappointing, however, as the power output has been much lower than originally expected. The reasons for this appear to be due to several factors, including reduced wave energy reaching the Limpet device and inefficiencies in conversion of the wave energy to pneumatic power and in the air turbine

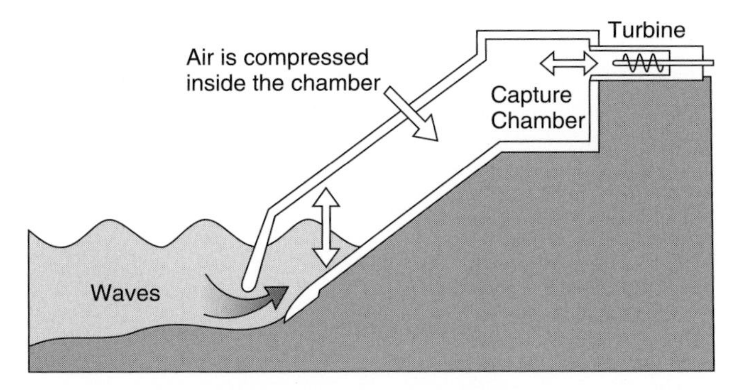

Figure 7.8 Schematic of "Limpet" oscillating water column device.

itself. Excavation and construction of the main concrete structure had to be carried out behind a temporary cofferdam, which meant that the finished structure was located some 15 m inland from the coastline. This resulted in a substantial attenuation of the incoming wave power from an estimated 20 kW/m at the coastline to 12 kW/m at the actual Limpet location. Operating experience also revealed that the Wells turbine, operating in a reversing airflow, was much less efficient than originally envisaged, with a measured efficiency of only 40%. The final measured power output of the Limpet was 21 kWe, compared with an initial design estimate of some 200 kWe. This demonstration has shown that although ocean waves contain large amounts of energy, the conversion into useful electrical power is challenging and expensive.

Many different types of off-shore wave energy devices have been proposed, and most of these rely on the use of wave action to provide a partial rotary motion to devices such as the Salter "Nodding Duck," or to convert the simple heaving motion imparted to a floating device as input ultimately to an electrical generator. A sketch of the Salter Nodding Duck device is shown in Figure 7.9. In this concept, developed at the University of Edinburgh, passing waves impart a partial rotary motion to a cam-like device (the "Duck"), which is free to rotate around a central shaft anchored to the sea bed. The oscillating rotary motion can then operate a hydraulic pump or some other device to provide the power to drive an electrical generator. These types of device have the advantage of access to the full off-shore wave energy potential, without the attenuation experienced by on-shore devices. They suffer, however, from the challenge of constructing very large devices that can withstand severe storms, and from the need both to generate power in difficult conditions at sea, and to transmit this power back to shore.

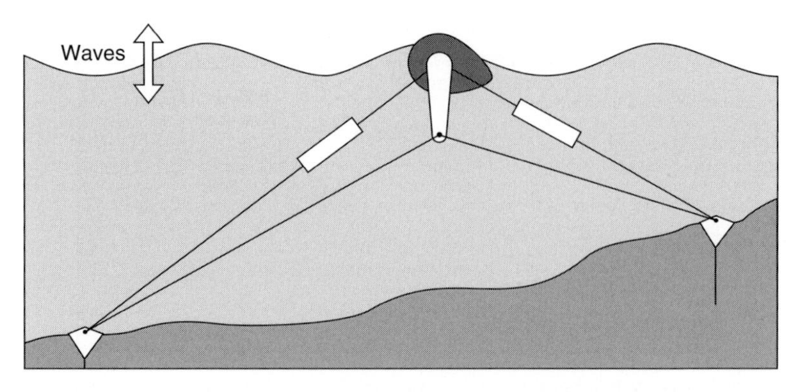

Figure 7.9 The Salter "Nodding Duck" wave energy device.

Another consideration, given that a large surface area needs to be covered in order to generate a significant amount of power, is the potential hazard to shipping. Many researchers continue to work on a wide range of devices, however, and further development work will undoubtedly result in optimization of some of them so that the unit cost of electricity from such devices can be made more competitive with traditional sources.

7.7 GEOTHERMAL ENERGY

Geothermal energy is the only renewable energy source other than tidal power that does not depend on the sun as its primary energy source. The high temperatures that prevail deep in the earth's crust have been recognized for a long time as a significant potential source of energy, both for space heating and for the generation of electricity. The use of geothermal energy is most practical in regions where the ground temperature is high close to the earth's surface, and these are often adjacent to geologically active areas that provide natural hot springs or geysers. This source of energy has been used by mankind since ancient times, usually in the form of natural thermal baths, but the search for alternatives to fossil fuels has led to renewed interest in geothermal activity. Most geothermal energy is used directly to provide heat for buildings and industrial processes, and by the end of 2000 the world-wide installed thermal capacity for non-electric heating applications was over 15 000 MWt (International Geothermal Association, 2005). Iceland is the third largest user of geothermal energy for heating following the USA and China, with some 1470 MWt of heating capacity used in 2000. This is expected to grow substantially in the coming years, however, as Iceland positions itself to be a leader in the use of renewable energy. In some countries, notably the USA, the Philippines, Mexico, and Italy, geothermal energy is also a significant source of primary energy for electricity production. At the end of 2003 the worldwide geothermal electricity generation capacity was some 8400 MWe, with the USA leading the way with 2020 MWe of installed capacity, closely followed by the Philippines with 1930 MWe. For the USA, however, the geothermal capacity provides less than 0.5% of total electrical energy generation, while for the Philippines it represents nearly 22% of total generation. Iceland had a much smaller installed geothermal electricity capacity of around 200 MWe, but since it is a very small country this accounts for nearly 15% of total electricity generation.

The generation of electricity is accomplished using conventional steam powerplant technology, but the overall system design varies considerably, depending on the type of geothermal energy source being accessed. The simplest type of powerplant feeds the "dry steam" produced naturally at some geothermal sites directly to a steam turbine, which provides the power to drive a generator, just as in a conventional fossil-fueled powerplant. Steam is formed naturally when ground water encounters hot rock at depths up to a few kilometers below the earth's surface. The steam generated in this way can sometimes find its way to the surface through natural fissures in the surrounding rock, as evidenced by geyser activity such as those made famous by the regular eruptions of the "Old Faithful" geyser in Yellowstone Park in Wyoming, USA. In most cases, however, the steam formed at depth does not naturally reach the surface, but can be readily accessed by drilling wells into a geothermal "reservoir." The temperature and pressure of this naturally generated steam is usually much lower than for a conventional fossil-fuel powerplant, resulting in lower overall thermal efficiency and the need to use turbines specially designed for these conditions. The very first experiments to see if geothermal energy, in the form of dry steam, could be used directly to generate electricity were undertaken at Larderello, Italy, in 1904. It began with tests using a small reciprocating steam engine of a few kWe capacity, and the success of those experiments led to steady expansion so that the installed capacity now at Larderello is some 550 MWe. The only other dry steam powerplant in existence today, and the world's largest geothermal powerplant, is "The Geysers" plant located in northern California. This plant began operation in 1960 with an initial capacity of 11 MWe, and today has an installed capacity of nearly 1700 MWe.

Not many geothermal sources produce dry steam at a temperature and pressure suitable for direct use in a steam turbine, however, and for lower temperature sources, which usually consist of high pressure hot water, or a mixture of saturated water and vapor (usually called "wet steam"), a so-called "flash-steam" approach is used. In this design the hot liquid or natural wet steam source is fed into a vessel which is held at a much lower pressure so that the liquid "flashes" into vapor which is then fed into a low-pressure steam turbine. Depending on the pressure and temperature of the incoming water or wet steam source, some liquid may remain at the lower pressure, and this is separated in the vessel and returned to the earth in a re-injection well. Most geothermal powerplants operate in this manner, and usually make use of both production wells and re-injection wells so that there is minimal environmental effect. For even lower temperature sources a "binary cycle" powerplant design is sometimes used. In this design a heat exchanger is used to transfer heat from the hot water exiting the production well to a secondary fluid, usually a refrigerant or other low-boiling-point fluid, so that vapor can be generated and used to drive a turbine. In this case the refrigerant or "binary fluid" is condensed after exiting the turbine and then recycled back through the heat exchanger in a closed loop. There is then complete separation of the working fluid used in the binary cycle from the geothermal source, and the cooled effluent water is sent to a re-injection well immediately upon exiting the heat exchanger. The more specialized equipment required for a binary cycle plant, together with the lower operating temperature, results in a higher capital cost compared with a simple dry steam plant. In the past this higher capital cost has somewhat limited the expansion of low-temperature geothermal electricity generation, but increased prices for fossil fuels and the need for sustainable energy sources will no doubt lead to a significant expansion of geothermal power production in the coming decades.

The use of "ground-source heat pumps" provides a way of obtaining significant amounts of thermal energy from very low temperature geothermal sources, or even from subsoil a few meters below the earth's surface. Because the ground temperature remains quite constant just below the earth's surface, this can be used as a source of heat in most parts of the world. A heat pump, working like a refrigerator in reverse, takes in energy from the ground at a relatively low temperature, and then delivers it at a higher temperature, usually for use in space heating applications. In this type of installation a pipe loop is buried in the ground near the building to be heated, and refrigerant from the evaporator side of the heat pump is then circulated through this loop. The ground heat is used to evaporate the refrigerant, which is then compressed to a higher pressure and temperature before being piped to the condenser. The condenser is a heat exchanger which then transfers heat to a building heating system as the refrigerant is cooled and converted back to liquid form. Electrical energy is required to drive the heat pump, of course, but with a "coefficient of performance" of the heat pump greater than unity, this is a much more efficient use of electricity for heating than using electric resistance heating. The capital cost of the heat pump system is significantly higher than a simple resistance heating system, but again, as energy costs increase this capital cost can usually be offset by reduced operating costs. An added benefit of a heat pump system for building heating is that the heat pump can be run in reverse during the summer cooling season, and can therefore provide both summer air conditioning as well as winter heating. This feature can make heat pumps an attractive building HVAC (Heating, Ventilating, and Air Conditioning) choice in regions with large temperature changes from winter to summer.

BIBLIOGRAPHY

Berndes, *et al.* (2003). The contribution of biomass in the future global energy supply: a review of 17 studies. *Biomass and Energy*, **25**, 1–28.

Boyle, G., *et al.* (2004). *Renewable Energy – Power for a Sustainable Future*, 2nd edn. Oxford: Oxford University Press.

Clark, R. (1995). Tidal power. In *Encyclopedia of Energy Technology and the Environment*. New York: John Wiley and Sons.

Enercon (2005). *http://www.enercon.de/en/_home.htm*

European Wind Energy Association (2005). *http://www.ewea.org/*

Frau, J. P. (1993). Tidal energy: Promising projects: LaRance, a successful industrial scale experiment. *Energy Conversion, IEEE Transactions*, **8** (3), 552–8.

Grubb, M. J. (1986). The integration and analysis of intermittent sources on electricity supply systems. Ph.D. thesis, Cambridge University.

Huttrer, G. W. (2001). The status of world geothermal power generation 1995–2000. *Geothermics*, **30**, 1–27.

International Energy Agency (2005). *http://www.oja-services.nl/iea-pvps/pv/home.htm*

International Geothermal Association (2005). *http://iga.igg.cnr.it/geo/geoenergy.php*

Maine Solar House (2005). *http://www.solarhouse.com/*

Marine Current Turbines (2005). *http://www.marineturbines.com/home.htm*

Middelgrunden Wind Farm (2005). *http://www.middelgrunden.dk/*

National Renewable Energy Laboratory (2005). *http://www.nrel.gov/*

Northern Ireland Assembly (2005). *http://www.gov.uk/enterprise/reports/report3-01rvol1.htm*

Pimentel, D. and Patzek, W. (2005). Ethanol production using corn, switchgrass, and wood; Biodiesel production using soybean and sunflower. *Natural Resources Research*, **14** (1), 65–76.

Sheehan, J., Aden, A., Paustian, K., Kendirck, K., Brenner, J., Walsh, M. and Nelson, R. (2004). Energy and environmental aspects of using corn stover for fuel ethanol. *Journal of Industrial Ecology*, **7** (3–4), 117–46.

Sheehan, J., Camobreco, V., Duffield, J., Graboski, M. and Shapouri, H. *An Overview of Biodiesel and Petroleum Diesel Life Cycles*. NREL Report: NREL/Tp-580-24772.

Tucson Electric Power (2005). *http://www.tucsonelectric.com/*

University of Strathclyde, Energy Systems Research Unit (2005). *http://www.esru.strath.ac.uk*

US Department of Energy (2005). *http://www.energy.gov/*

8

Nuclear power

8.1 INTRODUCTION

The inclusion of nuclear power in a section on "New and sustainable energy sources" may seem controversial to some readers. However, nuclear energy is today an important primary energy source which produces no greenhouse gas emissions while generating electricity. In fact, in some countries nuclear power provides a significant share of electrical power generation, and accounts for nearly 80% of all electrical power production in France, for example. Nuclear power was originally developed in the 1950s for the peaceful application of the very large quantities of energy released by the splitting of atoms, or "nuclear fission," and by 2001 it accounted for 17% of all electricity produced worldwide. The very first nuclear station to generate electricity began operation in Russia in 1954, with a capacity of just 5 MWe. The first commercial-scale nuclear powerplant, however, was the Calder Hall station, opened in the UK in 1956, consisting of four reactors each with an electrical generating capacity of 50 MWe. During the early years of nuclear power development it seemed that this source would provide an inexhaustible source of low-cost electricity, and it was pursued aggressively in much of the developed world. After considerable expansion through the 1960s and 1970s, significant cost overruns and two serious nuclear power accidents in the 1980s brought about a change in the public perception of the safety, security, and cost of nuclear power. This resulted in a dramatic reduction in the construction of new plants in most parts of the world. In recent years, however, the growing realization that the use of fossil fuels for electricity generation may be an important contributor to global warming has led many countries to re-evaluate the role that nuclear power may play in the quest for reduced greenhouse gas production. After first surveying the current status of nuclear power technology we will return to this issue, and the public perception and acceptance of nuclear power, at the end of this chapter.

Natural uranium, as found in nature, normally consists of about 99.3% U^{238} with the remaining 0.7% being a "fissionable" isotope, U^{235}. Nuclear power production harnesses the very large amount of thermal energy, or heat, which is released during a nuclear fission reaction when U^{235} absorbs a neutron and is split into fission products after being bombarded by a stream of neutrons. The amount of thermal energy released from just one kilogram of U^{235} undergoing fission is equivalent to that obtained by burning some 2.5 million kilograms, or 2500 tonnes, of coal. One of the attractions of nuclear power is this extremely high energy density of the nuclear "fuel," which greatly reduces the mass of material needed to generate electricity. In its natural form U^{235} is quite unstable, and a small fraction of the material may spontaneously undergo a fission reaction, the result of which is a number of fission products and one or more neutrons. The neutrons produced, however, are so-called "fast neutrons," which pass right through most of the U^{235} without being absorbed and causing further fission reactions to take place. Only if these fast neutrons are slowed down, by some form of "moderator," are they able to consistently trigger most of the U^{235} to undergo a fission reaction, leading to a sustainable chain reaction and the production of heat. If a moderator is present and a chain reaction is sustained, then more neutrons are produced than are absorbed, and a great deal of heat is released by the continuous fission of U^{235}, and this can then be used to generate steam. Several different substances are effective in acting as a moderator, and slowing down the fast neutrons to create the chain reaction, but the most commonly used are ordinary water, graphite, and "heavy water." Their use will be explained in more detail in the following sections describing the major types of commercial reactors currently used for nuclear power production.

8.2 LIGHT-WATER REACTORS

Most nuclear powerplants operating today make use of so-called "light-water" reactors as a source of heat for generating steam to drive conventional steam turbine generators. These are referred to as light-water reactors, mainly to distinguish them from "heavy-water" reactors, and use ordinary water both as a moderator and as a coolant to remove heat and produce steam. Heavy water is water containing deuterium, a hydrogen isotope with a neutron as well as a proton in its nucleus, rather than hydrogen, which has just one proton in its nucleus. This very special form of water has the ability to be a very effective moderator of the nuclear fission reaction, and will be discussed in more detail in the next section. Because ordinary water, or "light water," has a relatively poor ability to moderate the nuclear fission reaction, the fissionable uranium used as "fuel" in light-water reactors must be enriched to increase the concentration of the fissionable isotope U^{235}. In practice a light-water nuclear reactor "core" consists of enriched fuel contained in a series of fuel rods which are then surrounded by ordinary, or "light" water which acts as a moderator. The presence of the moderator surrounding the fuel rods enables a sustained fission reaction to take place, resulting in the generation of large quantities of heat. If the core were to be left uncooled for any length of time this heat would very quickly increase the temperature to such an extent that the fuel rods would melt. This is prevented, however, by circulating cooling water through the core, which is then used to generate steam for use in a steam turbine. In some cases the same water is used as both coolant and moderator in a direct-cooled reactor, while in others there are separate water supplies for both coolant and moderator in an indirectly cooled reactor design.

The simplest type of light-water reactor is the direct-cooled "Boiling Water Reactor," or BWR in which the same water is used as moderator and coolant, and as steam to drive the turbine generator. A schematic of this type of plant is shown in Figure 8.1 (US Nuclear Regulatory Commission, 2006). In this configuration water from the powerplant condenser is pumped into the reactor vessel, a large steel pressure vessel, by the feed pumps. As the water passes through the reactor core some of it boils, and the steam vapor formed then rises to the top of the vessel where it is fed back to the steam turbine generator where it provides the driving force to produce electricity. To ensure good circulation of the water around the core and adequate cooling of the fuel rods, a number of circulating pumps are also provided. A number of control rods passing through the bottom of the reactor vessel and entering the core can also be seen. These are made of materials such as cadmium or boron which strongly absorb neutrons, and can be used to control the degree of reactivity by moving them in or out of the core. Moving the control rods out of the core is equivalent to increasing the firing rate in a conventional fossil-fueled plant, while moving them completely into the core causes the fission reaction to be completely stopped. The reactor vessel and auxiliary equipment like the recirculation pumps and control rod mechanism is usually

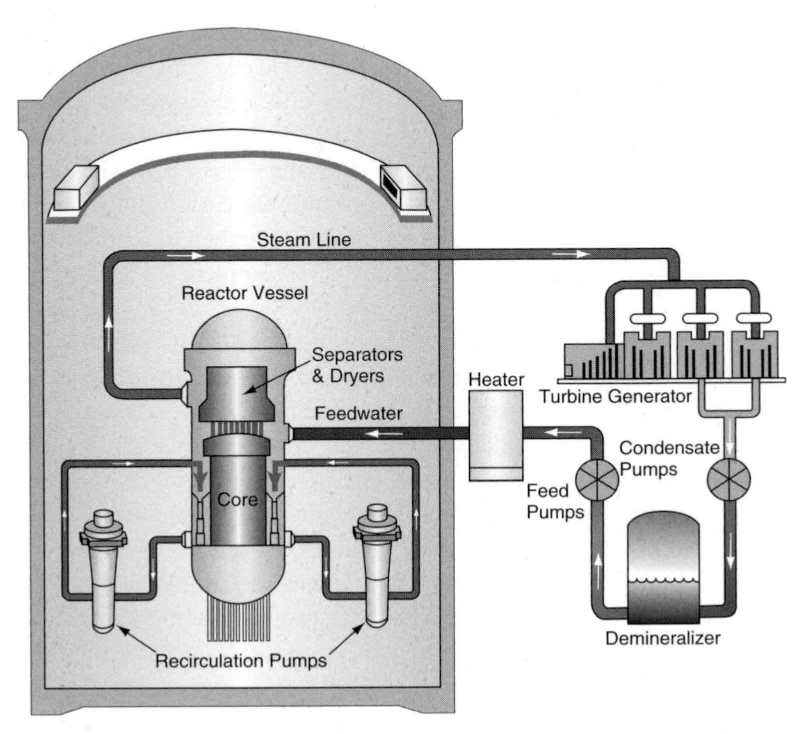

Figure 8.1 Boiling water reactor. *Source*: US Nuclear Regulatory Commission.

contained in a thick-walled concrete building which acts to absorb any excess radiation which may pass through the walls of the reactor vessel itself. Although BWRs are in principle very simple, the change in phase of the water from liquid to vapor within the reactor core provides some control challenges, making this reactor type less common than its cousin, the indirectly cooled Pressurized Water Reactor, or PWR.

A schematic of a PWR reactor is shown in Figure 8.2 (US DOE-EIA, 2005), which shows that two separate water circuits are used in this type of design, so that the heat generated in the reactor core is transferred indirectly to the steam circuit used to drive the turbine generator. In the first, or primary coolant circuit, water under high pressure is continuously circulated through the reactor core by a number of coolant pumps. This primary coolant, which also acts as a moderator, is kept at sufficiently high pressure so that it never boils at the temperatures reached in the reactor vessel. In this way the control problems associated with water undergoing a phase change from liquid to vapor are avoided, resulting in a simplified control system compared

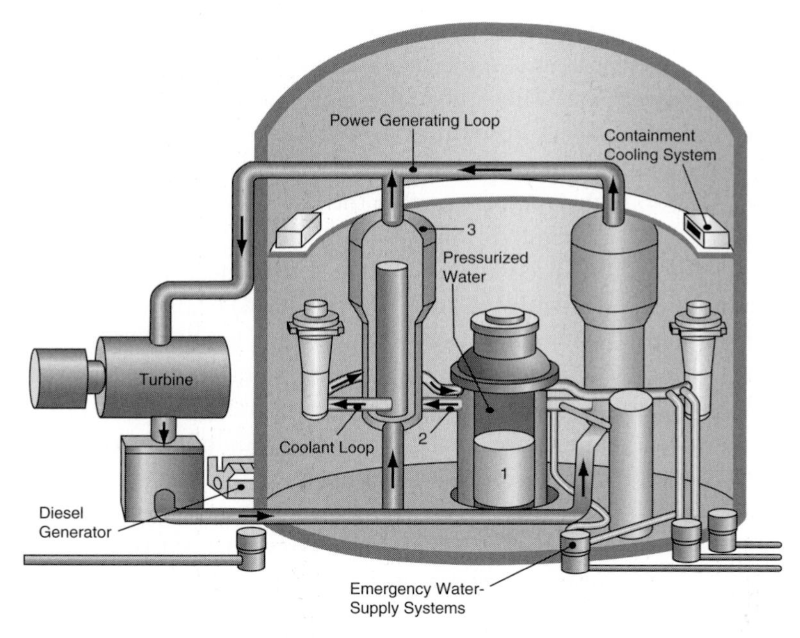

Figure 8.2 Pressurized water reactor. *Source*: DOE-EIA.

with that needed for a BWR design. The coolant pumps continuously circulate the primary coolant water through the reactor core and out into a series of heat exchangers which are used to transfer heat from the primary coolant to a secondary coolant water system which is maintained at a lower pressure, and therefore boils, providing a constant supply of steam to drive the steam turbine generator. This separation of primary and secondary coolants has proven to be a very effective design, and the PWR is the most common type of nuclear reactor powerplant in use today. It is the most common type of reactor used in the USA, the world's largest generator of nuclear electricity, and is exclusively used in France, the world's second largest producer. In fact, of approximately 440 power generation reactors operating in the world today, more than half are PWR designs.

One of the factors common to both the BWR and PWR designs is the need for very large and strong pressure vessels. These are critical for the containment of the reactor core, and also need to withstand the high temperatures and radiation fluxes inherent in generating high power levels. The reactor vessel can be nearly 5 m in diameter and 15 m high, with a wall thickness of more than 20 cm (8 inches). Any crack or fracture in the reactor vessel and the associated piping can lead to a loss of coolant and the ability to remove heat from the reactor core, which is one of the most important safety concerns with operation of a nuclear powerplant. There are only a few facilities in the world where such large steel structures can be constructed to the very high standards required to ensure the integrity and safety of the reactor core. Also, in both types of reactors the use of ordinary water as a moderator, which tends to absorb many of the neutrons released during the fission reaction, means that a sustained chain reaction cannot be obtained with the small fraction (0.7%) of fissionable U^{235} which is normally contained in natural uranium. In order to provide a sufficient supply of neutrons to sustain a fission reaction the uranium fuel needs to be enriched so that the concentration of U^{235} in the uranium fuel is increased to be within a range of approximately 3–5%. The enrichment process used to increase the concentration of fissionable U^{235} is in principle very simple, but because there are such small physical and chemical differences between the two uranium isotopes this becomes a very complex process in practice. The separation process needs to separate the two isotopes based solely on the small difference in the number of neutrons contained in their nucleus, which results in a very small change in their respective atomic masses. Enrichment can be done using either gaseous diffusion techniques, or a gas centrifuge approach. In the gaseous diffusion process a feedstock of uranium hexafluoride is first converted into the gas phase by heating, and then fed through a series of special porous membranes which preferentially pass the lighter U^{235} isotope. The gas centrifuge process uses a large number of high-speed centrifuges which preferentially feed the heavier U^{238} isotope towards the outside of a container, where it can be extracted to leave the remaining gas enriched in U^{235}. Only a very few countries have such enrichment facilities, and because of the ability of enriched uranium to be used for production of nuclear weapons these facilities are closely monitored by the international community.

8.3 HEAVY-WATER REACTORS

If a moderator that is much more efficient than ordinary water in enabling the fission reaction is used, then natural uranium, rather than enriched material, can be made to sustain a fission reaction and may therefore be used to fuel a nuclear reactor. Heavy water, or deuterium oxide, is water in which the usual hydrogen atom is replaced by a deuterium atom, containing one neutron as well as one proton in its nucleus. The advantage of using this form of water as a moderator for a nuclear powerplant is that it no longer has the ability to absorb stray neutrons, unlike ordinary water. Also, the presence of an extra neutron in the heavy water molecule acts to slow down "fast neutrons" produced in the fission reaction, so that they can trigger a sustainable chain reaction using only natural uranium fuel containing just 0.7% U^{235}. Heavy-water reactors therefore use natural uranium fuel and heavy water as the moderator, and sometimes also as the primary coolant. Although there is no longer the need for complex uranium enrichment facilities, this is somewhat counter-balanced by the need to produce heavy water, which is done by increasing the concentration of the small quantities of deuterium oxide that are naturally found in ordinary water using a combination of chemical and physical processes.

Heavy-water nuclear powerplants have been extensively developed in Canada, using the "CANDU" (Canadian Deuterium Uranium) design. These plants are different in two major respects from the light-water reactors used in most other countries. The first, and obvious difference, is the use of heavy water for both moderator and primary coolant, which enables the use of natural uranium, produced in large quantities in Canada, as a fuel. The second, and less well-known difference is that CANDU reactors use a "pressure tube" reactor core design, rather than the "pressure vessel" design used in light-water reactors. The advantage of this design is that the high pressure coolant, as well as the nuclear fuel, is contained in a series of relatively small (10 cm diameter) tubes rather than in one large pressure vessel. Because they are much smaller in diameter than a single large reactor vessel, the pressure tubes can be much thinner (approximately 5 mm) rather than requiring the 20 cm or more wall thickness used in a single large vessel. These tubes can be readily manufactured in most parts of the world, so that the construction of a CANDU plant is not dependent on the ability to manufacture very large pressure vessels, which exists in only a few countries. A schematic of a CANDU nuclear powerplant is shown in Figure 8.3 (AECL, 2005). The pressure tubes can be seen in a horizontal position passing through a "calandria" vessel which is filled with heavy water to act as moderator for the natural uranium fuel. The fuel is contained in a series of "bundles" which are placed inside each of the pressure tubes, and these are designed so that the primary coolant (also heavy water) can circulate through the bundles as it is pumped through the pressure tubes by the heat transport pumps. The primary heavy water coolant picks up heat from the fission reaction as it passes through the pressure containment tubes, and is then fed through a

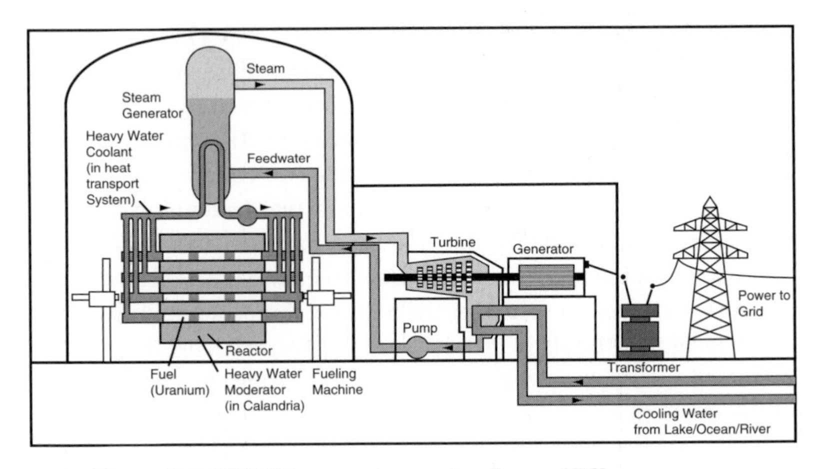

Figure 8.3 CANDU heavy-water reactor. *Source*: AECL.

series of loops in the steam generator vessel. The steam generator is simply a heat exchanger which transfers heat from the circulating heavy water primary coolant to the ordinary water secondary coolant which then boils to produce steam to drive the turbine generator. The schematic also shows another unique feature of the CANDU system, which is the automatic refueling machine placed at each end of the calandria vessel. Because the natural uranium fuel is depleted of fissionable U^{235} much sooner than in a reactor using enriched fuel it needs to be replaced more frequently. The refueling machines do this automatically, while the reactor is operating, which eliminates the need to shut-down the reactor for re-fueling, as is done with light-water reactors.

8.4 OTHER REACTOR TYPES

In the UK the first generation of commercial nuclear powerplants, such as those at Calder Hall, used gas rather than water as the primary coolant. This design uses large blowers to circulate CO_2 through the reactor core to remove the heat and then transfers this heat to ordinary water in a series of heat exchangers, or steam generators. They are usually referred to as "Magnox" reactors, since the uranium fuel is encased in a magnesium oxide casing. Graphite is used as the moderator, which permits the use of natural uranium fuel. This reactor type was superseded by the AGR (Advanced Gas Cooled Reactor) for a second generation of gas-cooled reactors in the UK. The AGR also uses carbon dioxide as a coolant, and a graphite moderator, but uses enriched uranium fuel and operates at higher temperatures and pressures so that steam conditions are comparable to those in a fossil-fueled plant. Although the AGR reactors have proven to be safe and operate with a high efficiency, they have suffered from some reliability problems and are quite costly to operate. The last nuclear powerplant to be built in the UK, Sizewell "B," was of a more conventional PWR light-water cooled design. No nuclear plants have been built in the UK since Sizewell "B" opened in 1995, although the British government has indicated in recent energy policy statements that it would "... keep the nuclear option open." It is unclear, however, if any new nuclear powerplants built would be gas cooled, or would be of the more conventional PWR or BWR designs used in much of the rest of the world.

In the former Soviet Union block of countries, principally Russia, two different types of light-water cooled reactor design have been used. The first design, referred to as RBMK reactors, uses a series of pressure tubes for containment, similar to the Canadian CANDU design, but these are orientated in a vertical position through a graphite moderator block, and light water is used as the coolant. These are also directly cooled reactors, similar to the BWR design, in which the primary coolant water is converted into steam within the reactor, and then fed directly to the turbine generator unit. One weakness with this design is that with a fixed level of moderation provided by the graphite, which is also flammable, excess steam formation in the core can reduce the ability to remove heat without reducing the intensity of the fission reactions taking place. This can then lead to an unstable operating regime for the plant, which was unfortunately demonstrated in the worst possible way during the serious fire and core meltdown of one of the Chernobyl reactors. This incident in 1986 has been well documented as the worst nuclear accident in history, resulting in loss of life and putting a nearly complete halt to further construction of RBMK reactors. The second type of reactor used in Russian nuclear powerplants is referred to as VVER, and is a light-water cooled and moderated design very similar to the PWR reactor used in Western countries. It evolved from the reactors used to power nuclear submarines, and like all PWR reactors is inherently much safer than the RBMK design.

Breeder reactors, as the name implies, are used to produce additional sources of fissionable nuclear fuel. We have already noted that natural uranium consists of 99.3% U^{238} with the remaining 0.7% being the fissionable U^{235} isotope. However, during the operation of a nuclear reactor the high neutron flux results in some of the normally "wasted" U^{238} being converted to a plutonium isotope, Pu^{239}, which can readily undergo a fission reaction. The Pu^{239} is usually treated as a waste-product, albeit a highly radioactive one which must be handled carefully. If the spent fuel is reprocessed, however, the plutonium can be separated from the rest of the fission products and can be used subsequently as fuel. Only small amounts of this man-made fissionable material is created in conventional power reactors since most of the neutrons produced are "slow" neutrons because of the presence of the moderator. A special breeder reactor, without a moderator present, and using enriched uranium fuel, can be used to produce much larger quantities of fissionable Pu^{239}. These reactors can then produce much more fuel than they consume, and the Pu^{239} can be stockpiled for use as fuel in conventional power reactors. In this way, a much greater fraction of the natural uranium energy source could be utilized than is currently possible in conventional reactors. Estimates have indicated that up to one-half of the uranium contained in natural uranium could ultimately be used as nuclear fuel in this way, rather than only 0.3% as at present. This much better utilization of the natural uranium would have the effect of expanding the availability of nuclear fuel supplies by a factor of more than 100. Although several countries have operated experimental breeder reactors, no commercial-scale breeder reactor programs have yet been implemented.

8.5 ADVANCED REACTOR DESIGNS

It is generally considered that the commercial development of nuclear power from its inception in the 1950s until today has gone through two generations of design. The first generation were really demonstration plants, aimed at proving that nuclear power was a viable commercial technology for electricity generation. These first-generation plants included the Magnox gas-cooled reactors in the UK, and first generations of PWR designs from Westinghouse in the USA and Framatome in France, and early BWR reactor designs from GE in the USA. Many of these first-generation units had modest power output of less than 100 MWe, and most have now reached the end of their useful design life, and many have been shut down as a result. The second generation of nuclear powerplants were full-scale commercial plants, usually with a unit-size of 500 MWe, or greater, and these have proven to be the backbone of the nuclear industry until today. Many of these plants are also now reaching the end of their design life, and the major developers of nuclear power have developed a new generation of both light- and heavy-water reactors. This second generation of reactors tended to be one-off designs, with each new station incorporating some new design features as a result of experience gained in the operation of previous units. This evolutionary practice, with many design changes being introduced during construction, tended to increase the capital cost of the second-generation plants. The new "Generation III" designs which have been proposed all use modular design to keep capital costs down, and have also incorporated a number of new design features aimed at enhancing safety and reliability. The unit sizes have also increased, and are now in a range from 600 MWe to 1600 MWe for the largest units.

In the USA, Westinghouse (Westinghouse Electric Company, 2005) has developed new modular PWR designs for both 600 MWe and 1000 MWe plants, and these are referred to as the AP600 and AP1000 designs. Also in the USA, General Electric (General Electric Company, 2005) has developed a 1350 MWe Advanced Boiling Water Reactor (ABWR), and most recently a so-called "Generation III+" design, the "Economic Simplified Boiling Water Reactor" (ESBWR). This 1500 MWe design provides further simplification of the ABWR concept, and incorporates "passive safety" features by designing for natural circulation of the cooling water through the reactor core. In this way, in the event of a complete station power failure, no pumps would be required to provide sufficient water circulation to control the core temperature so that the prospect of a core meltdown would essentially be eliminated. The ESBWR design requires many fewer pumps and valves than previous generation designs, resulting in a smaller plant footprint and reduced cost. All of these attributes are attractive to potential customers, and several utilities have expressed interest in using the ESBWR for new powerplants, assuming all of the regulatory requirements are successfully completed.

In both the advanced PWR and BWR reactor concepts substantial modularization of the design has resulted in improved safety and reduced capital costs. One of the other advantages of this type of modular design and construction is a reduction in the time required to construct a nuclear plant to between 3 and 4 years, which is much shorter than was required for Generation II plants. This shorter construction time has a major impact on reducing capital costs, with estimates for both GE and Westinghouse light-water reactors in the range of $1400 to $1600 per installed KWe. In Europe the generation III EPR 1600 (European Pressure Reactor) is a 1600 MWe design (Framatome, 2005) developed jointly by a consortium of Framatome and Siemens. This is an evolution of the PWR units previously constructed in France and Germany, with the emphasis again being on modular construction in order to improve safety and reduce construction time and capital costs. Olkiliuto 3 in Finland is the first EPR 1600 plant to be ordered, and is now under construction, with a planned start-up date in 2009. Not to be outdone by their light-water competitors, AECL in Canada (Atomic Energy of Canada, 2005) have been working to develop a new Generation III version of the CANDU pressurized heavy-water reactor. The ACR (Advanced Candu Reactor) 700 is a 700 MWe design, and an even larger 1000 MWe unit, the ACR 1000, is also being proposed. In addition to taking advantage of a modular design and construction approach, the ACR units have also incorporated a number of design changes aimed at increasing performance and reducing costs. These include the use of slightly enriched uranium fuel so that light water rather than heavy water can be used as a primary coolant, which has reduced the required heavy water inventory by some 75% (heavy water is still used as a moderator, however). The new design also incorporates higher steam pressure and temperature conditions which results in higher thermal efficiency with a consequent reduction in operating costs.

There are also a number of so-called "Generation IV" reactor designs in the very early stages of development. Most of these are conceptually quite different from previous generations, and have so far been mainly design exercises. Nevertheless they promise considerable advantages in terms of increased efficiency and reduced operating costs, and some may be ready for commercial development within the next 20–30 years. The US Department of Energy has led an international consortium of 10 countries in the Generation IV International Forum (or GIF), and this group has identified six advanced reactor types which could be developed for commercial operation about 30 years from now (US DOE, 2005). The goal of this group is to develop a more sustainable nuclear energy supply which would make much better use of available uranium supplies, and provide increased safety and reliability, together with reduced production of nuclear wastes and lower costs for power production. The six reactor types identified by the GIF in their "roadmap" document are mostly "fast" reactor designs which provide much better fuel utilization and use novel primary coolants suitable for high temperature systems. They include both "once-through" reactor designs, which rely solely on fission of U^{235}, and "closed cycle" designs which recycle plutonium generated in the fast reactors. The Gas Cooled Fast Reactor System (GFR) utilizes a closed fuel cycle with recycling of the plutonium generated to greatly reduce the requirement for uranium. It would use helium as a coolant to directly drive a gas turbine operating with an inlet temperature of around 850 °C to generate electricity. The Lead Cooled Fast Reactor (LFR) also uses a closed fuel cycle, but is essentially a "sealed" design, requiring refueling only after 10–30 years. The lead coolant circulates by natural convection at a relatively low temperature of around 600 °C. The LFR is envisaged for distributed power generation with unit sizes ranging from 50 to 150 MWe. The Molten Salt Reactor (MSR) operates with a liquid fuel formed from a mixture of sodium, zirconium, and uranium fluoride, which is circulated through a graphite moderator core. The MSR operates with a low primary coolant pressure of around 5 bar which contributes to safe operation, and the heat is then transferred to a secondary coolant for utilization in a conventional Rankine cycle power generation system. Another fast reactor is the Sodium Cooled Fast Reactor (SFR), which uses metallic fuel but is cooled by liquid sodium circulating at a relatively modest temperature of approximately 550 °C. The final two systems considered in the GIF study were the Supercritical Water Cooled Reactor (SCWR) and the Very High Temperature Reactor (VHTR), which uses helium as the primary coolant. The SCWR could operate on either an open or closed fuel cycle, and would use water above the critical pressure and temperature to achieve a compact and efficient design. The VHTR would, like the GFR, use helium gas to drive a gas turbine, but in this case a temperature close to 1000 °C would enhance the overall plant efficiency. A variant on the VHTR is the so-called "pebble-bed" reactor system, which would use spherical fuel pellets arranged in a relatively simple core design. Most of these designs only exist as paper studies to date, and would need considerable design and development work before they would be ready to enter commercial service.

And, finally, we should mention nuclear fusion as a potential long-term alternative to the conventional fission reactors currently in use and planned for the next several decades. In a fusion reaction the nuclei of hydrogen isotopes, like deuterium and tritium, are fused together under tremendous pressures and temperatures of millions of degrees to produce a helium atom and a high-energy neutron, thus releasing large amounts of energy. It is this energy source that provides the sun's vast energy output, and scientists have dreamed for decades of harnessing the fusion reaction for terrestrial use. The potential benefits of developing nuclear fusion powerplants on earth include an essentially inexhaustible form of "fuel" in the form of hydrogen isotopes extracted from seawater, an increase in the safety of nuclear powerplants, and a significant reduction in the quantity of radioactive waste materials that would be produced. Fusion energy has been studied in the laboratory for many years, but the engineering challenges of constructing a practical reactor capable of containing plasmas at temperatures many orders of magnitude greater than the melting point of any known material for more than a fraction of a second have so far been insurmountable. To date scientists have tried using either magnetic confinement in which an extremely strong magnetic field is used to contain the reacting nuclei, or inertial confinement provided by focusing two very powerful laser beams on a small target of "fuel." Magnetic confinement, in the form of a "Tokamak" device was used to produce a nuclear fusion reaction for a fraction of a second in 1997, but more energy was needed to power the device than was produced by the fusion reaction. Also, the energy required for the very powerful lasers proposed for inertial confinement would be many times greater than that produced by the resulting fusion reaction. The elusive goal, therefore, is to develop a technique to provide a self-sustaining fusion reaction with a net positive energy output, which can then be used to generate a continuous source of electricity or another energy carrier such as hydrogen. The scientific and engineering challenges of doing this are incredibly difficult, and the successful demonstration of a nuclear fusion powerplant may not occur for some 50–100 years, if ever. For the time being, therefore, it would not be prudent to rely on the development of nuclear fusion as a practical source of power in the foreseeable future.

8.6 NUCLEAR POWER AND SUSTAINABILITY

As we mentioned in the introduction to this chapter, some readers may question the choice of nuclear power as a sustainable energy source. However, the concept of "sustainability" is a relative one, and as a result of the second law of thermodynamics, which states that the total entropy in the universe is always increasing, there is no such thing as "complete sustainability." We have chosen to focus on nuclear power in this chapter since its use to generate electrical power, in place of fossil fuels, is clearly one of the ways to reduce the generation of greenhouse gases. Also, even though uranium availability is not limitless, just like any other natural resource, the use of "breeder" type reactors could extend the availability of nuclear fuel for hundreds of years. In this section, therefore, we will examine the major sustainability issues which could affect plans in several countries for an expansion of nuclear power as a way to move towards a more sustainable energy supply. These issues include the availability of uranium as the principal natural fuel for nuclear plants, as well as safety and nuclear proliferation, nuclear waste storage, and the economics of nuclear power production. Finally, we will examine the issue of public acceptance of nuclear power and look at France as a "case study" of the use of nuclear power to reduce the demand for fossil fuels.

The currently known "proven reserves" of uranium, of about 3–4 million tonnes, are sufficient to provide fuel for all of the approximately 440 existing nuclear powerplants for about the next 50 years. It has been estimated that with a doubling in the price of uranium, however, this could be increased by a factor of about 10 (US DOE, 2005). The assessment of 50 years of supply at today's prices assumes that the uranium fuel is used in the current generation of nuclear powerplants in a "once through" or open-cycle mode in which case only some 0.7% of the natural uranium which is in the form of fissionable U^{235} is used. In other words, over 99% of the uranium is not utilized, and ends up as part of the waste stream which must be stored or disposed of in some way. In the longer term, however, a transition to some of the "Generation IV" technologies could be made, in which fast reactors would be used to convert a substantial portion of the otherwise unused U^{238} into the plutonium isotope Pu^{239} and other actinides capable of undergoing nuclear fission reactions. The use of such "closed cycle" reactor designs could then extend the availability of nuclear fuel by a factor of more than 50 times. This, together with the expected increase in proven uranium reserves, would result in a large increase in the amount of fissionable material available for power generation. Even with a substantial expansion of nuclear power use there would then be nuclear fuel availability for more than 1000 years, which would meet many people's definition of sustainability. Ultimately, of course, even these supplies would be insufficient for the very long term, but by that time there may well be a long-term transition to renewable energy as the ultimate source of sustainable energy.

The safety of nuclear powerplants is, rightly, of great concern to both the general public and to the operators of these plants. There have been two serious accidents at operating nuclear powerplants, both resulting in extensive damage to the nuclear core in what is usually termed a "loss of coolant" accident. The loss of coolant from any nuclear reactor removes not only the ability to maintain the temperature of the reactor core within design limits, but also usually removes one of the major control elements needed to sustain stable operation.

The first major incident occurred in 1979 at the Three Mile Island plant, operated by Metropolitan Edison in Harrisburg, Pennsylvania (Merilo, 1980). The extensive damage to the reactor core in this case came about as the result of a classic case of a series of relatively minor mechanical failures, compounded by inappropriate operator response. Unit 2 at Three Mile Island was a conventional PWR design in which high-pressure water used as a primary coolant is circulated through the reactor core to remove heat, and then transfers this heat to a secondary coolant in a series of heat exchangers, or "steam generators," used to produce the steam needed to drive the turbine generator. In light-water reactors, such as the PWR design, both the primary and secondary coolants are just ordinary water. The primary coolant pressure is maintained at approximately 150 bar (2200 psi) so that the water does not boil, thereby maintaining its effectiveness as both a coolant and a moderator, while the secondary coolant is maintained at a lower pressure so that it does boil to produce steam suitable for driving the turbine. If, during operation of the reactor there is any significant loss of either the primary or secondary coolant, this greatly reduces the ability to remove heat from the reactor core, necessitating a rapid but controlled shut-down of the reactor. All reactor designs incorporate a sophisticated safety monitoring and control system which is designed to provide automatic reactor shut-down in the event of any major loss of coolant.

On the morning of March 28, 1979, some routine maintenance work was being done on a condensate polisher unit which removes impurities from condensed steam so that it can be re-used in the plant. A maintenance worker was using a compressed air line to attempt to clear a blockage in a small water line. As a result, water was forced back into the air system past a leaking check valve, and eventually entered the instrument air system which is used to operate many valves. Due to this contamination of the instrument air system, a number of unplanned control actions took place, leading the reactor safety system to begin automatically shutting down the reactor. These procedures included starting a series of emergency feedwater pumps which are designed to provide an additional source of heat removal in order to maintain the reactor core temperature within normal limits. Usually, this would have ended like many other minor incidents, with a complete plant shut-down, controlled automatically by the reactor safety system. In this case, however, a series of "block" valves had been inadvertently left closed, so that the emergency feedwater was not able to be pumped into the secondary coolant system. Due to the lack of coolant on the secondary side there was no way to remove heat from the primary coolant, and its temperature, and therefore pressure, increased to a level that caused a safety valve to open. Unfortunately, this valve failed to close again once the primary coolant pressure was reduced, leading to the loss of most of the primary coolant. Also, since the primary coolant had started to boil, the pressure remained high, leading the operators to mistakenly believe that there was much more primary coolant in the system than was actually the case. They therefore intervened manually to reduce the flow of emergency coolant which had been started automatically, and this, together with the failure of the pressure relief valve to close, led to a partial meltdown of the reactor core in a classic "loss-of-coolant" accident. Fortunately, there was no loss of life as a result of the accident at Three Mile Island, and many lessons were learned both about "fail-safe" design, and about operator training. As a result, the new Generation III reactor designs are inherently much safer than the earlier designs, and operator training has been expanded to ensure that safety systems are not interrupted during automatic shut-down procedures. The most important lesson learned is probably that the design of complex engineering systems can be made safe from either mechanical failure, or human operator error, but usually not from a combination of both.

The second, and much more serious, accident occurred in 1986 at Chernobyl, near Kiev in the Ukraine, formerly part of the USSR. The reactor at Chernobyl was an RBMK type, which is a boiling water reactor, but uses graphite as the moderator, rather than water as in the BWR reactor design used in other countries. This type of design, however, can result in unstable operation during some operating conditions, since the moderator function is separated from the coolant function. The BWR type of reactor is a "once-through" design, in which steam is generated directly in the reactor core, unlike a PWR system in which high-pressure water is maintained always in the liquid phase as the primary coolant, and then transfers heat to a secondary coolant, as we have seen in the Three Mile Island plant. One potential problem with the BWR design is that excess steam formation in the core can reduce the ability to remove heat because of the higher fraction of vapor present. In a conventional water-moderated BWR a higher fraction of water vapor in the core also reduces the effectiveness of the water as a moderator, so that reduced core cooling coincides with reduced power output, resulting in stable operation. In the RBMK design, however, excess steam formation is not accompanied by a reduction in reactor moderation, so that a reduction in cooling capacity and full moderator capacity can potentially lead to unstable operation.

The RBMK design relies on a series of control rods, which consist of neutron absorbing materials, to control reactivity and maintain reactor stability. During the period of April 25 and 26, 1986 the Chernobyl unit 4 reactor was, ironically, being run through a series of safety tests, in which fewer than the normal number of control rods were being used, and the plant's emergency cooling water supply was disabled. During one of these tests the reactor became very unstable and a spike in reactivity caused excessive steam production leading to a rupturing of the core. The core rupture resulted in the graphite moderator catching fire, and due to the lack of a heavy-walled containment building, highly radioactive gases were released and transported around the world by the prevailing winds.

Unfortunately, some 30 lives were lost, either as a direct result of the reactor explosion and fire at Chernobyl, or due to radiation poisoning shortly afterwards. Although there have been many studies of what the longer-term health effects of the Chernobyl accident have been, and there is clear evidence that thousands of people received excessive doses of radiation, there appears to be no consensus about how many fatalities may be ultimately linked to the disaster. An annex to the report by the United Nations Scientific Committee on the Effects of Atomic Radiation, released in 2000 and commenting on the long-term effects of the Chernobyl accident (UNSCEAR, 2000), concluded in part that:

> Finally, it should be emphasized that although those exposed as children and the emergency and recovery operation workers are at increased risk of radiation-induced effects, the vast majority of the population need not live in fear of serious health consequences from the Chernobyl accident. For the most part, they were exposed to radiation levels comparable to or a few times higher than the natural background levels, and future exposures are diminishing as the deposited radionuclides decay. Lives have been disrupted by the Chernobyl accident, but from the radiological point of view and based on the assessments of this Annex, generally positive prospects for the future health of most individuals should prevail.

There is no doubt that this major accident had worldwide repercussions for the nuclear industry as a whole, and not just for the USSR. It also led to renewed efforts to quantify the risks associated with nuclear power development, and ultimately to the adoption of safer design concepts and enhanced safety standards all over the world. As with any technology, however, there will inevitably be other accidents, and ultimately the risks of nuclear power development will have to be weighed up against the potential benefits to mankind of providing a large-scale source of reasonably priced electricity without the generation of greenhouse gases.

The prospect of nuclear proliferation, in which nuclear materials are used by hostile countries to develop nuclear weapons, or passed to terrorist organizations, also needs to be addressed thoroughly. These threats to society will also require a careful assessment of the trade-off between such possible misuse of nuclear material and the benefits of nuclear-generated electricity. We need to realize however, that "the genie is already out of the box," and that potentially dangerous nuclear materials will be available to someone who wants them badly enough, whether the peaceful use of nuclear energy expands or not. It seems likely that there will always be rogue elements who wish to use any technology as a means to force their will upon others, whether it is gunpowder, chemical poisons, or nuclear materials. In order to manufacture a nuclear warhead, access to highly enriched uranium and very complex manufacturing facilities are required, and these are unlikely to be obtainable even by sophisticated and determined terrorist groups. Nuclear "waste," however, may be obtainable at some point in the future, either from black-market sources, or perhaps even directly from hostile nations that have already developed a nuclear power capability. This material potentially could be used to build a so-called "dirty bomb," which would not actually result in a nuclear explosion, but would use conventional explosives to distribute radioactive material in a populated area. The results of such an attack could be similar to those from the release of highly toxic gases in confined areas, another potential terrorist threat requiring only relatively "low-tech" expertise. To safeguard against these possibilities we need to always be vigilant, and there needs to be a strengthening of international agencies such as the International Atomic Energy Agency, and an increase in international cooperation to ensure careful monitoring of all nuclear waste material so that technology which is potentially beneficial for many is not misused for the benefit of a few.

The long-term storage and/or disposal of nuclear waste materials is now being seriously addressed by many countries that operate nuclear powerplants. Because of the small quantity of nuclear "fuel" required to generate large amounts of electrical power, the quantity of waste generated is also very small. The spent fuel inventory from a 1000 MWe reactor operating for a year, for example, totals about 25–30 tonnes. To date, most of the waste material generated in commercial nuclear powerplants has been stored on-site at each station. When a nuclear reactor is re-fueled, usually during an annual plant shut-down, or continuously while the plant is "on-line" in the case of CANDU reactors, the spent nuclear fuel elements are removed and stored in an open tank of water similar to a large swimming pool. The water provides natural radiation shielding, and also serves to remove any heat which may still be generated by the used fuel elements. Although this type of on-site storage has been sufficient to contain nearly all of the waste produced during some 50 years of nuclear plant operation, a more permanent form of waste disposal is required. Nuclear waste materials are usually characterized as low-level, intermediate-level, or high-level wastes. The spent nuclear fuel leaving a reactor contains a wide range of radioactive materials, with half-lives (the time in which radioactivity drops to one-half its initial level) ranging from a few seconds to millions of years. In some cases, as we have seen, the spent fuel may be re-processed to recover fissionable material such as Pu^{239}, and in this case only some 3% of the original fuel remains as high-level waste containing highly radioactive fission products, some with very long half-lives. In this case the mostly liquid high-level waste is first evaporated, and the remaining solid material is then added to molten borosilicate glass and cooled into a solid glassy material. This vitrified waste is then ready for eventual containment in storage canisters for disposal or long-term storage. If no re-processing of the spent fuel is done, as is the case for all commercial nuclear waste in the USA and Canada, for example, then all of the spent fuel is treated as high-level waste, and must eventually be disposed of in what is known as the "direct disposal" option.

Over the years, many different techniques for the long-term storage of nuclear waste materials have been proposed. These include disposal in the deep ocean, underground retrievable storage in stable geologic formations, and even disposal by launching waste containers into deep space using well-established, but expensive, rockets or space vehicles. So far, the international consensus appears to be that the safest, and most economic, form of waste disposal will be the storage of high-level waste deep underground in stable geologic formations such as rock formations or salt deposits. Both the USA and Finland have announced plans for deep geologic disposal of high-level nuclear wastes. In 2001 the Finnish parliament approved the construction of an underground waste disposal facility at Eurajoki which will be capable of holding some 2500 tonnes of encapsulated nuclear wastes (World Nuclear Association, 2005). This is approximately the quantity of high-level wastes that will be produced by the four operating Finnish reactors over a 40-year period. The underground depository will be at a depth of some 500 m, and will be designed so that the waste canisters can be recovered at some time in the future if that is considered desirable. The location is currently undergoing a more thorough investigation, including testing and characterization of the rock material, and it is expected to be operational by 2020. The estimated long-term cost of waste management, including waste storage and plant decommissioning, has been estimated to be approximately 10% of the total cost of nuclear electricity generation. In the USA, the US DOE have selected Yucca Mountain in Nevada as the first permanent storage site for high-level nuclear wastes. In 2002 the US Congress approved the development of Yucca Mountain as a nuclear waste repository after some 20 years of scientific studies, and detailed engineering is now under way to develop the facility. This site was selected because of its remote location in a very dry and geologically stable region comprising primarily volcanic rock called "tuff" which was deposited some 12 million years ago. The waste material is to be stored in a series of tunnels some 200 to 500 m below the surface, and approximately 300 m above the water table. The underground repository will be designed so that tunnels can either be closed and permanently sealed, or left open to allow access to the waste by future generations if that seems desirable. The current timetable for the development of the Yucca Mountain nuclear waste repository indicates that the facility will be ready to accept nuclear waste in 2010 (US DOE-EIA, 2005).

8.7 NUCLEAR POWER ECONOMICS AND PUBLIC ACCEPTANCE

The cost of nuclear power is often raised as an impediment to expansion of its use for electricity generation. The major cost component of nuclear-generated electricity is the capital cost of the plant, since very little nuclear fuel is required, and its cost is relatively small. A comprehensive study of the costs of nuclear power, compared with both conventional coal-fired plants and natural gas-fired combined-cycle plants, was undertaken by an interdisciplinary group at MIT and published in 2003 (MIT, 2003). The base case assumptions in this study were capital costs of $1300/kWe for a new pulverized fuel (PF) coal-fired plant, $500/kWe for a combined-cycle gas turbine (CCGT) plant, and $2000/kWe for a light-water reactor (LWR) nuclear plant. The total O&M (operations and maintenance) cost of electricity from a modern nuclear plant, including fuel, was estimated to be about $0.015/kWh,

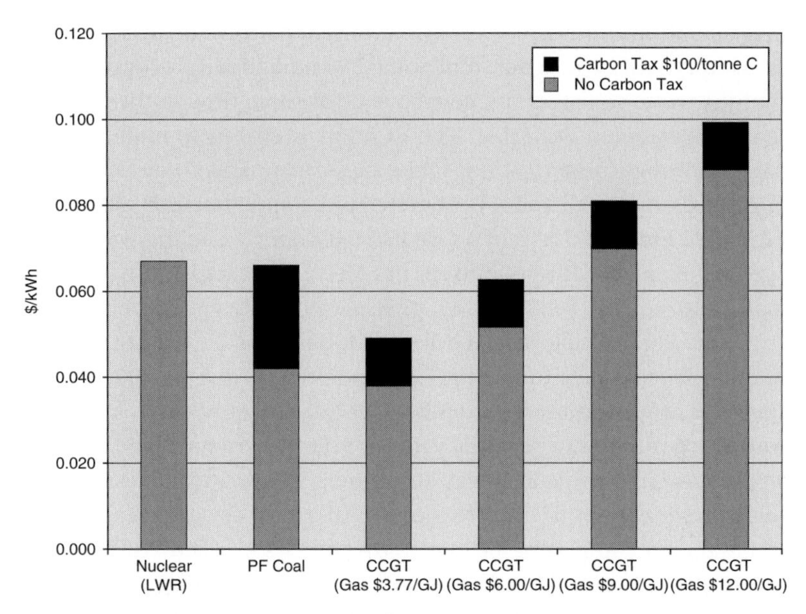

Figure 8.4 MIT estimate of electricity costs. *Source*: Massachusetts Institute of Technology (2003). *The Future of Nuclear Power: An Interdisciplinary MIT Study*. MIT Press.

while the fuel cost alone was estimated to be $1.20/GJ for the coal-fired PF plant. For the natural gas-fueled CCGT plant the cost of gas was assumed to range from a low of $3.77/GJ to a high of $6.72/GJ. The base case scenario assumed a 40-year operating life and an 85% capacity factor for all plants. These data resulted in an estimated total cost of electricity from the base-case nuclear powerplant of $0.067/kWh, while it was $0.042/kWh for the PF plant and a low of $0.038/kWh and a high of $0.056/kWh for the CCGT plant. Clearly, under these conditions the nuclear powerplant was not economically competitive with coal or natural gas-fueled CCGT plants, and illustrates why most powerplants ordered in recent years have been either coal- or gas-fired. However, since this study was published in 2003 the cost of natural gas has soared, with city-gate prices reaching as high as $12.00/GJ in 2005, and the environmental costs associated with the substantial greenhouse gas emissions from both coal- and gas-fired plants has come under increased scrutiny.

The results of the estimated cost of electricity from a nuclear plant compared with both a PF coal-fired plant and a natural gas CCGT plant from the MIT study are shown in Figure 8.4. For the CCGT cases, however, a range of gas costs is shown from the low-case of $3.77/GJ assumed in the study, to more realistic values representative of gas prices over the last 2 years. The electricity costs from a CCGT plant using these higher gas costs have been extrapolated from the costs assumed in the MIT study. Also shown in Figure 8.4 is the additional cost of electricity from all of the fossil fuel-fired plants if a carbon tax of $100/tonne of carbon were to be imposed. Assuming the imposition of this type of carbon tax, or some equivalent levy, the cost of coal-fired power becomes equivalent to that from a nuclear plant, while the cost of gas-fired power using natural gas prices experienced in 2005 becomes much more expensive than that from a nuclear powerplant. Under any similar scenario, with escalating fossil-fuel costs and increasing concerns about greenhouse gas emissions, nuclear power appears to be an attractive option. Ultimately, however, the decisions needed about whether or not to expand nuclear power production inevitably will be not only economic and technical, but also political in nature.

Public acceptance of nuclear power has certainly changed over the decades since it first appeared in the 1950s. In the beginning it was widely welcomed as a new, inexhaustible, and inexpensive source of electricity. Then, in the 1980s and 1990s, in the aftermath of the Three Mile Island and Chernobyl incidents, it fell into disfavor in most countries. This was compounded by the very large cost increases that often occurred as a result of on-going design changes necessitated by safety concerns raised during the design and construction of new plants during this time. Throughout the 1990s there was very little construction of new nuclear plants in the Western world, although new plants were planned and built in rapidly developing economies, such as China, India, and South Korea. Throughout the 50-year period since the first nuclear powerplant became operational, however, France has stood out as the one country in which there appears to be wide acceptance of nuclear power. While nuclear power today is the primary energy source for some 17% of the world's electricity generation, it accounts for nearly 80% of all the electricity generated in France, a far larger proportion than in any other country. This large-scale development of nuclear power has been continuous, and has occurred with seemingly little regard taken of the cessation of nuclear construction that has taken place in most of the rest of the Western world. The question naturally arises, therefore, as to why it should be that France should be so out of step with much of the world in terms of nuclear power development during the last two decades. Successive French governments have evidently chosen to invest in nuclear power, through the state-owned utility Electricité de France (EDF), partly because of concerns about security of energy supply and partly because of the environmental effects of burning fossil fuels to generate electricity. The widespread public acceptance of such a strategy, seemingly flying in the face of public reaction in most other Western countries, has been explained in the book *The Radiance of France* by the sociologist Gabrielle Hecht (Hecht, 1998). In this book, the author explains that the favorable opinion of nuclear power held by the French public is mostly due to the tight social structure of the country's corporate and bureaucratic elites. Many of the key decision-makers, both in the large industries that design and build nuclear power stations, and in the relevant government departments, are graduates of the elite universities in France, and the general public largely admires such people. This is the main reason, she argues, that nuclear power has been widely accepted, with very little dissension from either the public press or opposition parties. Whether this is an entirely accurate representation of the situation in France may be open to question, but it does illustrate the fact that public opinion on such important questions of energy policy is often shaped by factors other than purely technical and economic issues.

BIBLIOGRAPHY

Atomic Energy of Canada Ltd. (2005). *http://www.aecl.ca/*

Framatome (2005). *http://www.framatome.com/*

General Electric Company (2005). *http://www.gepower.com/prod_serv/products/nuclear/en/index.htm*

Hecht, G. (1998). *The Radiance of France: Nuclear Power and National Identity after World War II*. Boston, MA: MIT Press.

Hore-Lacy, I. (2003). *Nuclear Electricity*, 7th edn. Uranium Information Centre and World Nuclear Association.

Massachusetts Institute of Technology (2003). *The Future of Nuclear Power: An Interdisciplinary MIT Study*. Boston, MA: MIT Press.

Merilo, M. (1980). *Up the Learning Curve for Reactor Safety: The Accident at Three Mile Island*. Presented at the 1st Annual Canadian Nuclear Society Conference, Montreal, June 1980.

UN Scientific Committee on the Effects of Atomic Radiation (2000). Report Annex J.

US Department of Energy (2002). *A Technology Roadmap for Generation IV Nuclear Energy Systems*. Report GIF-002-00.

US Department of Energy. Energy Information Agency (2005). *http://www.eia.doe.gov/fuelnuclear.html*

US Nuclear Regulatory Commission (2006). *http://www.nrc.gov*

Westinghouse Electric Company (2005). *http://www.ap1000.westinghousenuclear.com/*

World Nuclear Association (2005). Nuclear Energy in Finland. *http://www.world-nuclear.org/info/inf76.htm*

Part IV Towards a sustainable energy balance

9

The transportation challenge

9.1 TRANSPORTATION ENERGY USE

Transportation accounts for just over a quarter of the total global demand for energy, as we have seen in Chapter 4. With increasing "globalization" and rapidly increasing wealth in countries with large populations, such as China and India, the fraction of total energy resources devoted to transportation is likely to increase in this century. Transportation energy demand can be divided between transportation primarily aimed at moving people, and that aimed primarily at moving materials and supplies, or "goods." A further division of energy demand can also be made between the main transportation modes, i.e. travel by land, by sea, and by air. The split in global transportation energy demand by mode has been estimated by the World Energy Council (WEC), and is shown in Figure 9.1 (World Energy Council, 2005). These data include transportation of both goods and people worldwide in the year 1995. Almost 80% of the total demand for transportation results from road transport, with just under 50% of the total demand being used to provide personal transportation in light-duty vehicles. The remaining 20% of total transportation demand is split nearly equally between the air, rail, and marine transportation modes. Nearly all of the energy used for transportation is derived from crude oil in the form of gasoline and diesel fuel for road transport, jet fuel for air travel, and diesel fuel and heavy bunker oil for marine transportation. The only exception to this is the use of electricity for some rail transportation, primarily in regions with high population densities such as Europe and Japan. The WEC has also predicted that the total demand for transportation energy will grow by some 55% over the period from 1995 to 2020.

Liquid petroleum fuels are ideally suited to transportation applications because of their inherently high energy density, and the ease of

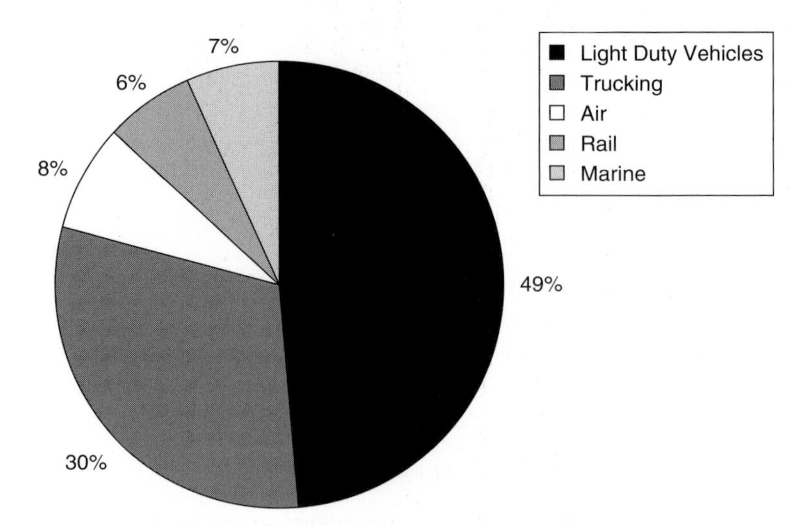

Figure 9.1 Global transportation energy demand by mode – 1995.
Source: Based on figures from the World Energy Council Statement *2000: Energy for Tomorrow's World – Acting Now!*

transportation and storage of these fuels. The internal combustion engine has reached a high level of development, and this is now almost universally used as the power source for all road vehicles. For aircraft, there is a need for as high a power to weight ratio as possible and the gas turbine engine, operating as either a pure jet engine or as a turboprop, is ideally suited to this application for all but the smallest aeroplanes. These engine families have been optimized to operate on petroleum-based fuels which are widely available in the form of gasoline, diesel fuel, and aviation jet fuel. The downside of using petroleum fuels, of course, is that they are all derived from crude oil, a non-renewable resource which will eventually be in short supply. Also, the combustion process produces emissions of nitrogen oxides, carbon monoxide, and unburned hydrocarbons, as well as large quantities of CO_2, the principal greenhouse gas. While we have discussed in Chapter 6 the prospects for dealing with CO_2 emissions using carbon capture and storage techniques, these are clearly not suitable for application to moving vehicles of any kind. The search is continuing, therefore, to find alternative energy sources for transportation, so that the very large contribution to greenhouse emissions from the transportation sector can be minimized.

One way to reduce the dependence of the transportation sector on petroleum-based fuels is to switch from the use of internal combustion engines fueled by petroleum to a completely different form of energy carrier. This has been done successfully for rail transportation by using electric locomotives on lines with heavy traffic volumes. This is possible for rail transportation since electrical power can be provided continuously to the locomotive through overhead electrical cables, or through a "third-rail" placed adjacent to the tracks. Although this provides a very clean source of energy at the point of end-use, if the electricity is generated primarily from fossil fuels, then there may be no net reduction of greenhouse gas emissions as a result of railway electrification. If, in the long term, the electricity carrier is generated primarily from non-fossil fuel sources, such as renewable energy or nuclear power, then there will be a direct benefit through the elimination of greenhouse gas production from the railways. However, even if all railways were electrified, and used non-fossil fuel primary energy sources, the contribution to reducing greenhouse gas emissions would be fairly modest since rail transportation accounts for only 6% of total transportation energy use, as seen in Figure 9.1. With transportation making up approximately 25% of total energy demand, this would then result in a reduction of just 1.5% in the global production of greenhouse gases. This is still a useful contribution, and we can expect continued progress on railway electrification, particularly where it can be justified by high passenger or freight load factors. For road vehicles, however, it is not practical to provide electrical power continuously to cars or trucks, and purely electric vehicles must rely on energy stored in an on-board battery. Although electric cars were common during the very early development of motor vehicles, the low energy capacity of batteries made them uncompetitive with vehicles powered by internal combustion engines, and they disappeared from the marketplace. Of course, during this time there was no consideration of the problems associated with the generation of greenhouse gases, and so the petroleum-fueled vehicle became the standard. As we shall see later in this chapter, however, recent concerns about greenhouse gas emissions, as well as about the long-term sustainability of fossil fuels, may well lead to a shift back to electric vehicles in many applications.

The use of liquid "biofuels," including ethanol and methanol as well as biodiesel fuel made from vegetable oils, is another way to combat the large contribution made by petroleum-based fuels to greenhouse gas emissions. Although these are still carbon-containing fuels, and therefore also produce CO_2 emissions, because they are derived from biomass sources, they are often considered to be "carbon neutral" since the biomass consumes CO_2 during the growing phase. The techniques for producing alcohol fuels are well-known, and they are suitable fuels for use in internal combustion engines, providing some minor modifications are made to the engine and fueling systems. Ethanol (the main ingredient in alcoholic beverages) is usually made from grain crops, and the land area required to produce the large quantities of ethanol required to satisfy all transportation requirements would be vast. Energy production in this way would likely compete directly with the land resources needed for food production, and as we have seen in Chapter 7 ethanol production may itself be an energy-intensive process. Methanol, manufactured from wood-waste, or from fast-growing tree species, could also become an important fuel source for transportation applications. These alcohol fuels have a high energy density, although only about one-half that of petroleum-based fuels, and their expanded use would require very little change in the infrastructure currently used to transport and store petroleum fuels. Although some changes to internal combustion engine design would be needed, these would be fairly minor, and could easily be phased into normal spark-ignition, or gasoline-fueled, engine production. In fact, both ethanol and methanol are currently used in some "blended" fuels, with gasoline being the main component of these fuels. Alcohol fuels are not, however, suitable for use in compression-ignition diesel engines, although biodiesel fuel could be manufactured from soybeans and other crops, as discussed in Chapter 7. In the long term, if petroleum-based hydrocarbons were found to be necessary for the production of aviation fuels, these could still be produced for many years from petroleum-like synthetic fuels derived from oil sands and oil shale, or even from coal using coal liquefaction techniques.

9.2 ROAD VEHICLES

For fueling road transportation in the future, particularly light-duty vehicles, there has been much speculation about the use of hydrogen as an energy carrier, which according to many authors and "futurists" would usher in the "hydrogen economy." Proponents of the hydrogen economy claim that the use of hydrogen as a transportation fuel would eliminate the production of any harmful exhaust emissions from vehicles on the road. This is, of course, true for the vehicle itself, but as we have noted in our discussions of the energy conversion chain in Chapter 2, it only represents one part of the complete energy use picture. Hydrogen would just be an energy carrier, like gasoline or electricity is today, and it would need to be "manufactured" from one

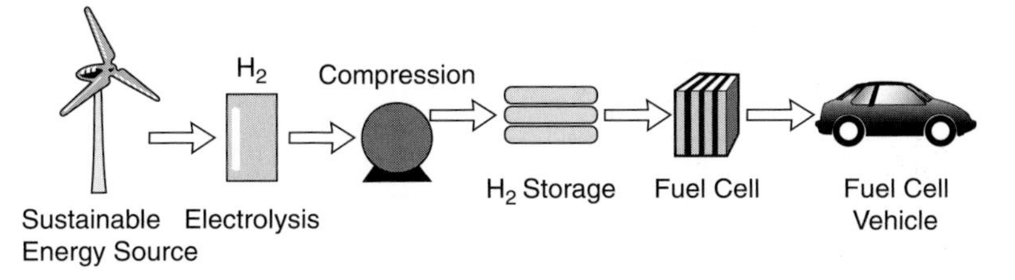

Figure 9.2 Energy conversion chain for a fuel-cell vehicle.

of the three primary energy sources. If this primary source were to be a hydrocarbon fuel, such as natural gas or coal, all of the carbon in the primary energy source would still end up as CO_2 at the point of hydrogen production. If, on the other hand, the hydrogen was produced from a more sustainable primary energy source, such as renewable energy or nuclear power, then there would indeed be no production of greenhouse gases anywhere in the energy conversion chain. The energy conversion chain for using hydrogen in this manner is illustrated by the schematic diagram in Figure 9.2. This shows the primary energy source being some form of sustainable energy, represented by wind power generating electricity in the figure, but this could be solar energy, or any other source of renewable energy or nuclear power. Following along the energy conversion chain, the electricity would then be used to produce hydrogen by electrolysis of water, and the hydrogen would then be compressed, or converted into liquid form, for storage on board the vehicle. The vehicle would utilize all-electric drive, and a fuel cell would be used to generate electricity on-demand from the hydrogen, which would then be supplied to an electric motor providing the mechanical power to drive the vehicle.

A fuel cell is, in principle, a very simple electrochemical energy conversion device which directly converts the chemical energy stored in a fuel like hydrogen, into electrical energy. The fuel cell was invented in 1839 by the Welsh inventor Sir William Grove, some 50 years before the internal combustion engine became a reality, and for the next 120 years was essentially a scientific curiosity. In the 1960s, practical fuel cells were developed for use on space vehicles in order to provide a steady source of electrical power to the spacecraft using liquid hydrogen, which was also used as a propulsion fuel. Then, in the 1970s a number of companies began to develop fuel cells for the production of electricity in place of conventional internal combustion engines or steam powerplants. There are several different types of fuel cell design, but all of them operate by first of all splitting hydrogen

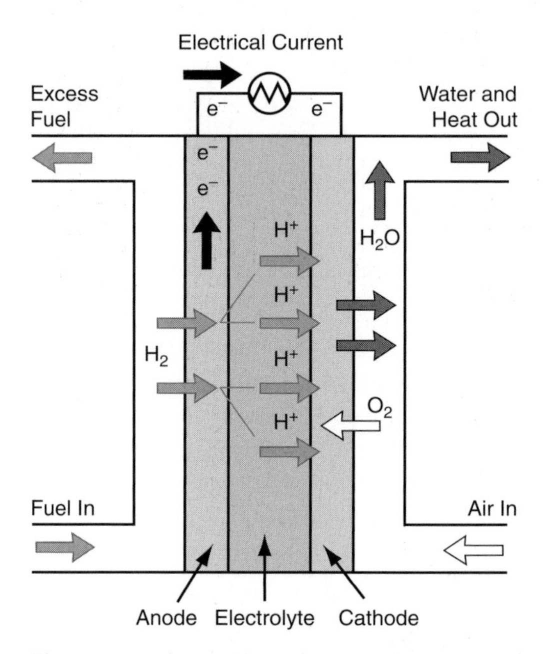

Figure 9.3 Polymer Electrolyte Membrane (PEM) fuel-cell operation.

atoms into a positively charged proton and a negatively charged electron, and then sending the electron through an external circuit to rejoin with the proton and an oxygen molecule to form water. If hydrogen and oxygen are continuously fed into the fuel cell, this process results in a stream of electrons, or an electric current, and this can then be fed into the electrical grid, or used directly to power an electric motor. Fuel cells are now also being pursued as a potential power source for motor vehicles, and the PEM ("Polymer Electrolyte Membrane," or "Proton Exchange Membrane") fuel cell is the usual design chosen for this application. The advantage of PEM fuel cells in this application is that they operate at relatively low temperatures (around 80 °C), and are quite compact compared with other designs, all of which are better suited to large-scale stationary power generation applications.

The operation of a PEM fuel cell is illustrated in Figure 9.3, which shows two plates, the anode and the cathode, separated by an electrolyte. Fuel, in the form of pure hydrogen, is continuously fed to the anode, while oxygen, or air, is fed to the cathode. The anode is coated with a noble-metal catalyst, usually platinum, which facilitates the ionization of the hydrogen atoms into separate streams of protons (shown as H^+) and electrons (shown as e^-). In this case the electrolyte is a solid polymer membrane, first developed by the du Pont Company under the trade-name "Nafion." The purpose of this non-conducting "proton exchange membrane" is to block the passage of electrons from the anode to the cathode, while at the same time permitting the protons to pass through. The electrons are then forced to pass through an external circuit, where they provide an electric current suitable for powering an electric motor, for example, on their way to the cathode. On the cathode the electrons then join with the protons which have passed through the electrolyte and oxygen atoms from the air, to form water molecules. The only other products from the fuel cell, in addition to the external electrical current, are water and heat. The figure shows just one "cell," which generates about 0.7 volts, while in practice many cells are placed together in a fuel cell "stack" so that the stack develops a higher overall voltage potential. Although very simple in concept, and with no moving parts, in reality a complete stand-alone fuel cell power unit becomes somewhat more complex with the need for compressors to overcome the pressure drop of fuel and air flows across the stack. The hydrogen fuel must also be very pure, and must not contain any trace of carbon monoxide, as this will quickly "poison" the catalyst, and prevent the efficient ionization of the hydrogen at the cathode. Polymer Electrolyte Membrane fuel cell stacks are also very expensive at the current state of development compared with internal combustion engines, for example, primarily due to the need for platinum catalyst material. More recent development is trying to reduce the cost of catalyst materials, however, and a mixture of platinum and ruthenium has been shown to be quite effective at much lower levels of catalyst loading.

If we now look back briefly at the energy conversion chain in Figure 9.2, we will see that the fuel cell is just one part of the hydrogen fuel-cell powered automobile. A very critical component of the vehicle propulsion system is the fuel storage system on board the vehicle. For a conventional motor vehicle, utilizing an internal combustion engine, this is the simple fuel tank, which stores either gasoline or diesel fuel, both of which conveniently exist as liquid fuels at normal ambient temperature and pressure conditions. At these same ambient temperature and pressure conditions, however, hydrogen is a gas, and this gas has a very low energy density, i.e., one cubic meter of hydrogen gas has a much lower energy content than one cubic meter of liquid fuel. In order to carry a significant quantity of energy on-board the vehicle in the form of hydrogen, therefore, it would need to be highly compressed, or perhaps even liquefied and stored in a "cryogenic" fuel

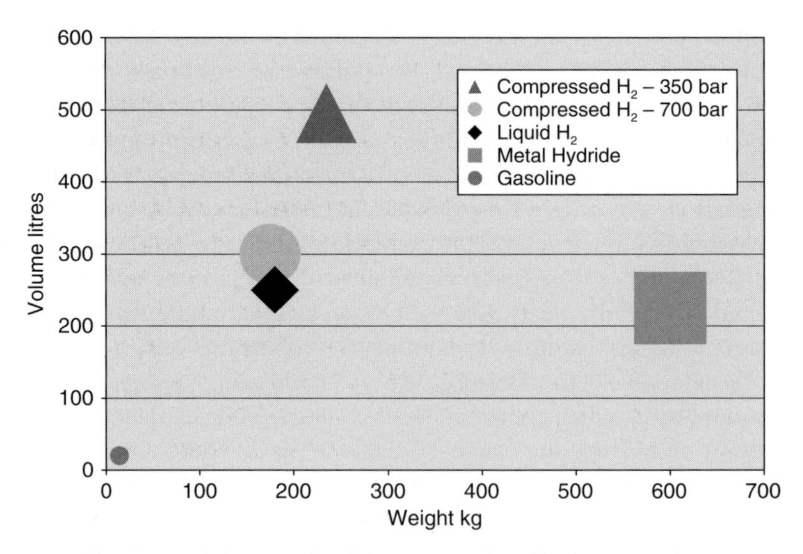

Figure 9.4 Volume and weight to store 5 kg of hydrogen, or its energy equivalent. *Source*: Rovera, G. (2001). Potential and limitations of fuel cell in comparison with internal combustion powertrains. Fiat Research. Presented at *ICE 2001*, Capri, Italy.

tank at a temperature of around −250 °C. In order to store enough hydrogen energy to provide a reasonable driving range, engineers have proposed using compressed hydrogen at a pressure of 350 bar (5000 psi), or even 700 bar (10 000 psi). These very high pressures require heavy gas storage cylinders, which would add considerable weight and volume to the vehicle compared with the usual sheet metal container used for liquid fuels. In fact, the storage of hydrogen on board vehicles is one of the most difficult challenges facing the successful commercialization of hydrogen-fueled vehicles.

This difficult hydrogen energy storage problem is summarized in Figure 9.4, using data from Rovera (2001), which plots the volume required to store a given quantity of energy against the total mass of the energy source and the container needed to store it. The data points on the diagram provide an estimate of the volume (shown on the left-hand vertical axis) and the total mass (shown on the lower horizontal axis) required to store 5 kg of hydrogen, or its energy equivalent. This amount of hydrogen would be the energy equivalent of about 18 litres of gasoline, or approximately one-half to one-third of the capacity of most car fuel tanks. If the hydrogen gas is stored in cylinders at a pressure of 350 bar, the volume of the containers would be approximately 500 litres, and their mass (plus 5 kg of hydrogen) would be on the order of 225 kg. This represents a volume larger than the total trunk space on most cars, and the mass would amount to about 25% of the total mass of a compact car. At a storage pressure of 700 bar the total storage volume would be reduced to around 300 litres, and the mass to about 180 kg. These numbers are not just one-half the values for storage at 350 bar, since cylinders with a much greater wall thickness would be required at the higher pressure. If the hydrogen were to be liquefied, and stored at ambient pressure in special cryogenic tanks at a temperature of −250 °C, then the mass of the container (and liquid hydrogen) would still be about the same as for high-pressure storage at 700 bar, and the volume only slightly less. Some proponents of hydrogen as an automotive fuel have suggested storage in the form of "metal hydrides." These are special alloys which catalyze the dissociation of hydrogen molecules at the metal surface, thus facilitating the absorption of hydrogen atoms directly into the metallic crystal lattice, enabling the storage of large volumes of hydrogen at ambient pressure conditions. However, this requires carrying a very large quantity of the absorbing metal, which would represent a mass of more than one-half that of a typical car, even when empty, as shown at the far right of the diagram. Finally, the small data point near the origin represents the volume and weight required to store 18 litres of gasoline, the energy equivalent of 5 kg of hydrogen, and clearly shows the benefit of the high energy storage capacity of conventional liquid hydrocarbon fuels. From these data it can be readily seen that the storage of hydrogen on board motor vehicles is likely to be one of the most important challenges facing anyone trying to commercialize hydrogen-fueled vehicles.

One of the advantages claimed for fuel-cell vehicles is the much higher energy conversion efficiency of fuel cells, compared with internal combustion engines. The efficiency of a PEM fuel cell, at around 50%, is certainly much higher than can be expected for the typical internal combustion engine used in motor vehicles. However, if the hydrogen used as the energy carrier on board the vehicle were to be derived from fossil fuels, as it will almost certainly be in any early stage of commercialization of such vehicles, then the overall "well-to-wheels" efficiency is unlikely to be significantly higher than that of the best available technology using a conventional internal combustion engine. This is the conclusion found in comparative studies of vehicle powertrain efficiency by both the Argonne National Laboratory of the US Department of Energy, and by researchers at the Massachusetts Institute of Technology. Figure 9.5 summarizes the

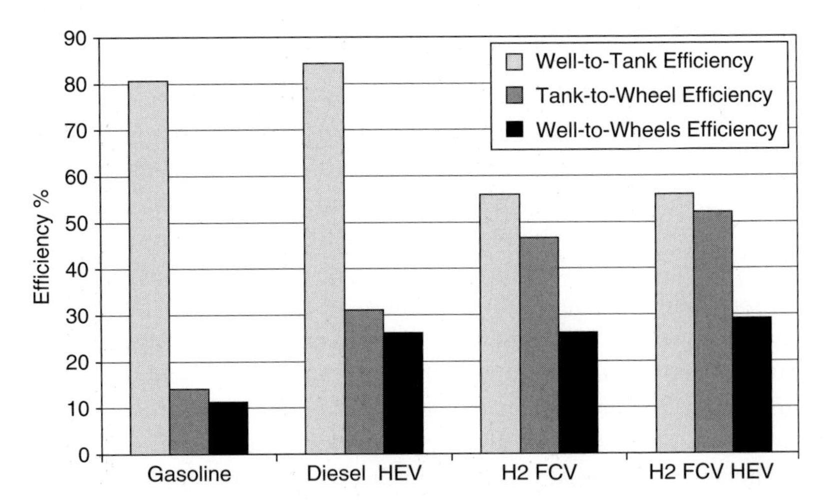

Figure 9.5 SUV well-to-wheels efficiency comparison. *Source*: Society of Automotive Engineers Technical Paper presented at the SAE 2003 World Congress & Exhibition, March 2003, Detroit, MI, USA, *Well-to-wheels analysis of advanced SUV fuel cell vehicles.*

results from the study published by the Argonne National Laboratory (US DOE-ANL, 2005). The study simulated the performance of several possible powertrain configurations for a Ford Explorer SUV, and the results for four of these are shown in Figure 9.5. The "prime movers," or power sources, used in each of the four cases shown were a conventional gasoline engine; a diesel engine operating in a hybrid electric vehicle (Diesel HEV); a hydrogen fuel-cell vehicle (H2 FCV); and a hydrogen fuel cell operating in a hybrid electric vehicle (H2 FCV HEV). For both of the fuel cell options shown, the hydrogen was assumed to be obtained from a refueling station by reforming natural gas, which would be the most likely source of primary energy, at least in the early phase of fuel cell commercialization. For each vehicle the efficiency of converting the primary energy, either crude oil or natural gas, into the on-board fuel, either gasoline, diesel fuel, or hydrogen, is shown as the "well-to-tank" efficiency. This can be seen to be just under 80% for gasoline, and just over that value for diesel fuel, while for obtaining hydrogen from natural gas the efficiency is approximately 56%. The efficiency of each prime mover was simulated while operating over a combination of the US Federal Highway Driving Cycle and the Federal Urban Driving Cycle, as recommended by the Society of Automotive Engineers. The results of this simulation are shown as the "tank-to-wheel" efficiency in Figure 9.5 for each case. Finally, the

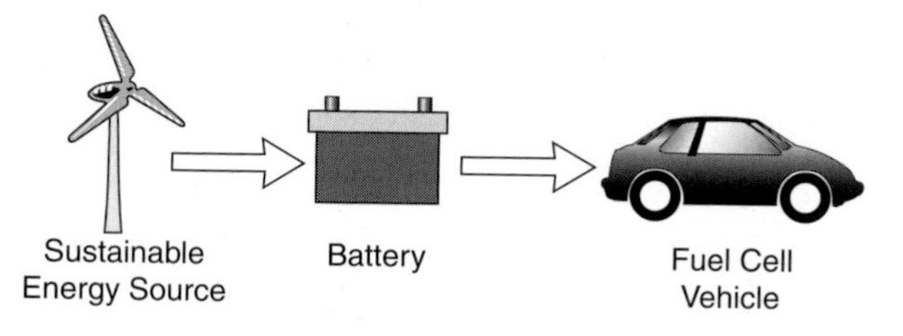

Figure 9.6 Energy conversion chain for a battery electric vehicle.

overall "well-to-wheels efficiency" is then found by multiplying these two efficiencies together. These results show that the overall "well-to-wheels" efficiency for the complete energy conversion chain is just about the same for the best internal combustion engine and hybrid vehicle configuration and for a simple hydrogen fuel-cell vehicle. For the case of a hydrogen fuel cell in a hybrid electric vehicle configuration it is just slightly greater at 29% than the diesel powered HEV at 26%. Also, if the primary energy source was a fossil fuel for both configurations, as assumed in the study, the result would be almost identical levels of CO_2 emissions. It seems unlikely, therefore, that this small gain in overall vehicle efficiency would be sufficient to overcome the much higher cost and complexity of the hydrogen storage system and the fuel cell itself. Advocates of fuel cell vehicles, however, contend that in the long term hydrogen will be produced from a more sustainable form of energy, perhaps renewable energy, as illustrated in Figure 9.2, or perhaps from nuclear power, which would then result in zero emissions of CO_2 for the complete energy conversion chain.

If we then consider the case in which hydrogen is generated from a renewable primary resource, or from nuclear power, we see from Figure 9.2 that the first step in the energy conversion chain is the generation of electricity as an initial energy carrier. This carrier is then converted into hydrogen as a secondary carrier, and this is stored on-board the vehicle. The final step in the chain is then the conversion of the stored hydrogen back into electricity by the fuel cell, and this is then used to power the electric propulsion motor. A parallel situation is used in a simple battery electric vehicle, in which a battery is used on-board the vehicle to store the electricity, as shown by the simple energy conversion chain schematic in Figure 9.6. In this case there is no need to convert the electricity into a secondary carrier, since the electricity generated as the primary carrier is stored directly by the battery, and then used when needed to supply the electric propulsion motor. The only difference between the energy conversion chains

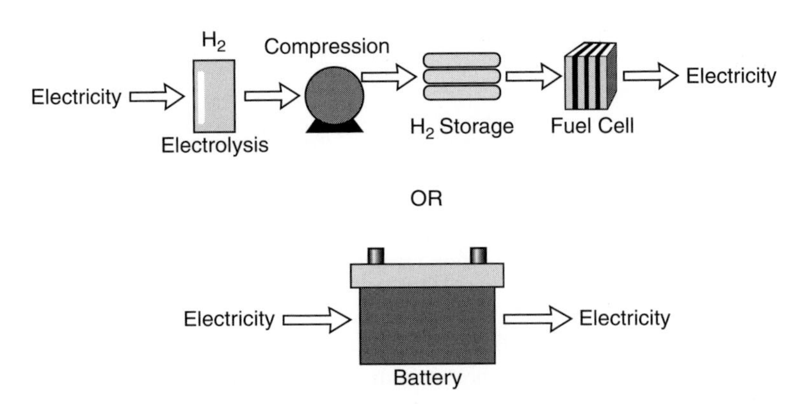

Figure 9.7 Alternative electrical energy storage concepts.

depicted in Figure 9.2 for the fuel-cell vehicle, and in Figure 9.6 for the battery electric vehicle, is that in the first case energy is stored in the form of hydrogen, and in the second case in the form of electrical energy in the battery. The difference in these two approaches can then be summarized graphically as shown in Figure 9.7. This shows the two different approaches, starting from the point at which the primary energy source produces electricity, and ending where electricity is again used to power the vehicle's electric traction motor. In other words, all of the equipment illustrated by the conversion chain in the top half of the figure, consisting of hydrogen production by electrolysis, compression and storage in high-pressure containers, and finally conversion back into electricity by a fuel cell, is directly analogous to the electrical battery in the lower half of the figure. By making a simple comparison of both parts of the figure, it can be seen clearly that all of the equipment required for the fuel-cell vehicle, including hydrogen production and storage and the fuel cell, is really just an electrical energy storage device. There is only an advantage of this approach over that of using a simple electrical storage battery, therefore, if the energy storage capacity on-board the vehicle is greater using the fuel-cell route.

Although strictly speaking not all of the process steps shown in Figure 9.7 are energy conversion processes, there is a loss of available energy associated with each step. For example, since it takes nearly 10% of the total energy content of the hydrogen to compress it to the pressure of some 350 bar used in the storage cylinders, for every 100 kJ of hydrogen energy produced by electrolysis, the net energy that then resides in the hydrogen storage is around 90 kJ. To account for these energy losses we may assign an "in-out" efficiency value to each step in the process, as

Table 9.1. *Electrical "in-out" efficiencies*

"In-Out" efficiency comparison			
Fuel cell		Battery	
Electrolysis	75%	Battery	90%
Compression	90%	–	–
Fuel cell	50%	–	–
Overall efficiency	34%	Overall efficiency	90%

shown in Table 9.1. For the battery, there is only one step between the electrical input to the energy storage and energy output to the vehicle, and since some of this energy is normally lost in the form of heat during the battery charging process, we can also assign an "in-out" efficiency to the battery. The charging efficiency is usually between 85% and 95% for lead-acid batteries, although it may be significantly lower for other battery types. If we assume a battery charging efficiency of 90%, however, the "in-out" efficiencies for each of the steps in the two equivalent energy storage processes of Figure 9.7 are shown in Table 9.1 for comparison. The efficiencies of each individual step are then multiplied together to get the final "overall" efficiency of the complete process, going from electricity "in" from the primary source to electricity "out" to the traction motor. With an assumed fuel-cell energy conversion efficiency of 50%, the overall "in-out" efficiency for the fuel-cell conversion chain shown in Figure 9.7 is approximately 34%, compared with 90% for the conventional battery. In other words, if we were going to use electricity from some sustainable primary energy source as the energy carrier for a vehicle, then for every 100 kJ of primary energy used, we would "deliver" some 90 kJ to the vehicle's electric traction motor using a lead-acid battery for energy storage. If, however, we used hydrogen and a fuel cell as the energy "storage" system, for every 100 kJ of primary energy used, only some 34 kJ would be delivered to the traction motor. We have used an estimate of the efficiency of commercial electrolytic production of hydrogen from water of 75%, although research and development is under way to significantly raise this value. Even if the efficiency of electrolysis were to be increased to 90%, however, the overall "in-out" efficiency would still only reach about 40%. This simple analysis indicates that an all-electric vehicle, using a high-efficiency battery with sufficient energy storage density to provide a reasonable vehicle range, would be a very attractive alternative to hydrogen storage and a fuel cell.

Table 9.2. *Energy storage density*

Energy carrier	Energy density MJ/l	Specific energy MJ/kg
Lead-acid battery	0.32	0.11
Advanced battery goal	1.08	0.72
H_2 gas, 200 bar	1.90	119.88
Gasoline	31.54	45.72

Source: US Department of Energy.

If electrical batteries were able to store sufficient energy to provide a range of up to 100 miles, then relatively simple battery electric vehicles, which would normally be re-charged overnight, or when not in use for several hours, would likely be attractive to most consumers. Such vehicles would be much less complex and much cheaper to produce than the comparable fuel-cell vehicles together with the necessary hydrogen production and storage systems. If these vehicles became the norm there would need to be an expansion in electricity capacity, although in the long term this could be a much more sustainable system than it is today. Table 9.2 shows the energy density (energy per unit volume) and specific energy (energy per unit mass), for lead-acid batteries and the goal under the US DOE Advanced Battery program (US DOE, 2005), as well as values for liquid gasoline and for hydrogen stored at a pressure of 200 bar (3000 psi). Although hydrogen has a very high specific energy, because it is a gas it has a very poor energy density, even when stored at high pressure. This means that very large (and therefore heavy) compressed gas cylinders must be used to store a significant quantity of energy. It can also be seen that batteries are not yet able to compete with liquid fuels in terms of either energy density or specific energy, and pure battery electric vehicles will likely be suitable only in specialized short-range applications for the foreseeable future. Much research and development work is being carried out on batteries, but there do not appear to be any major "breakthroughs" in battery design which would significantly increase the energy density to the point where simple battery electric vehicles are able to compete, in terms of range and performance with conventional vehicles. For the time being, therefore, battery electric vehicles are primarily used for low-speed off-road applications where range is not critical, in applications such as golf carts and motorized wheelchairs.

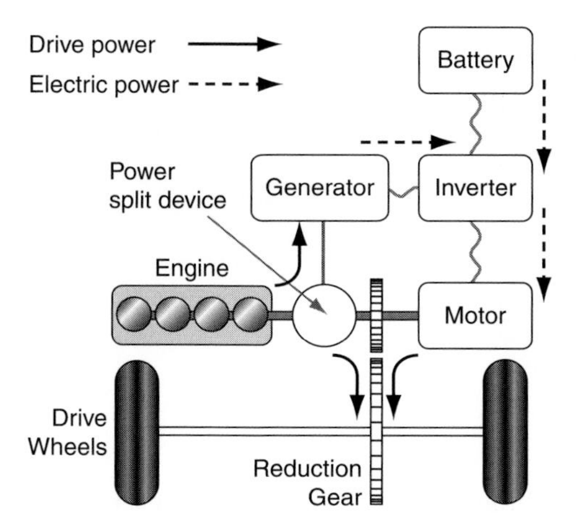

Figure 9.8 Operation of the series-parallel hybrid electric vehicle.

Much of the development work on batteries has been driven by the successful introduction in the last few years of hybrid electric vehicles (or HEVs). A simplified schematic diagram illustrating the operation of a hybrid vehicle is shown in Figure 9.8, which is based on the Toyota Prius design (Toyota Motor Corporation, 2005). This uses a propulsion system consisting of a conventional internal combustion engine, acting as the "prime mover," in parallel with an electric motor and storage battery. All of the energy to drive the vehicle still comes from the liquid fuel (gasoline or diesel fuel) used by the internal combustion engine, but the engine is used to either charge the battery via a generator, or to drive the wheels directly as in a conventional vehicle, or in some combination of both of these approaches. During low-speed operation, and particularly in stop-and-go driving in urban centers, the engine is shut down, and all propulsion is provided by the electric motor using electricity from the battery. As the battery becomes discharged the engine is automatically started and again begins to charge the battery, and may also provide some mechanical propulsion directly to the wheels through a gearbox. This type of "series-parallel" operation is made almost completely seamless to the driver by a control system which decides when to operate the engine, and when to shut it off, without any intervention from the driver. One major benefit of this powertrain design is that the engine can now operate at its most efficient design condition, independently of vehicle speed or load, thus greatly increasing the overall fuel efficiency. Another important feature of hybrid vehicles is the use of "regenerative braking," which utilizes the generator to absorb much of the braking energy normally

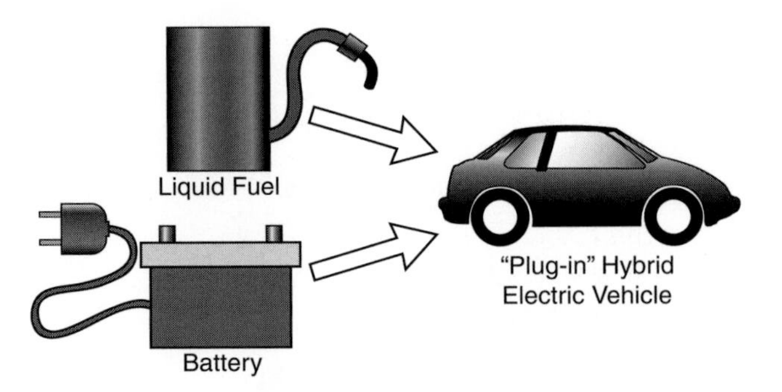

Figure 9.9 The "plug-in" hybrid electric vehicle.

dissipated in the form of heat by the brakes, and then uses this energy to recharge the battery. As a result of these design features a hybrid vehicle normally has better fuel mileage during city driving than on the highway, making them particularly well suited to urban commuting. These two features have been developed and refined by automotive engineers so that the hybrid vehicle has a fuel efficiency about 50% greater than that of a conventional vehicle powered by an internal combustion engine alone. Drivers also like driving hybrid vehicles because the electric motor is able to provide very strong acceleration due to its characteristic high torque at low speed. Hybrid vehicles have been very successfully introduced into the market, initially in compact cars, but the technology is now spreading to larger cars and sport utility vehicles where the benefit of much greater fuel economy will be particularly welcome.

The current design of hybrid vehicles may be classed as "grid-independent" or "stand-alone" hybrids, because they incorporate an electrical powertrain, and storage battery, but still obtain all of their primary energy from the fuel carried on-board the vehicle, and do not need to be plugged in to the electrical grid to recharge the battery. However, with the expected advances in battery energy density, and the desire to minimize the use of fossil fuels, these vehicles may very well be the precursor to a transition to the next generation of hybrid vehicles; the so-called "grid-connected" hybrids, sometimes also referred to as "plug-in" hybrids. A simple schematic of this concept, which incorporates the advantages of both pure battery electric vehicles and hybrid vehicles, is shown in Figure 9.9. In this concept, the battery pack in an otherwise conventional hybrid vehicle would be much larger, and could be fully charged when not in use by being plugged in to the electrical grid. The engine, however, would be smaller, and would still operate on some form of liquid fuel, as illustrated in Figure 9.9. In this way, the vehicle could operate for a significant range, perhaps somewhere between 20 and 60 miles (approximately 30–100 km), as a completely electric vehicle, and would use the engine to recharge the battery only when it was necessary to exceed this distance or perhaps when climbing very steep hills. For many drivers, and certainly for most commuters, the vehicle would then be capable of operating as a pure battery electric vehicle for most trips, and would be plugged in overnight and perhaps also when not in use during the working day. The successful development and introduction into the marketplace of the "plug-in" hybrid vehicle would mark the beginnings of a significant new transportation paradigm, that of disconnecting road vehicles from the need to use petroleum fuels, at least for the majority of miles traveled. In considering the complete energy conversion chain for this option, if electricity were to be generated primarily by sustainable primary energy sources, such as renewable energy or nuclear power, then road transportation would also become sustainable and would no longer be a significant factor in contributing to greenhouse gas production.

The Electric Power Research Institute in the USA has published the results of a study (EPRI, 2001) comparing the performance of two plug-in hybrid electric vehicle designs to a stand-alone HEV and a conventional vehicle powered by a gasoline engine. The study group modeled the performance of a typical mid-size automobile using the vehicle simulation program "ADVISOR" developed by the National Renewable Energy Laboratory of the US Department of Energy. The four different configurations were referred to as a conventional vehicle (CV); a stand-alone HEV with no all-electric range (HEV 0); and two plug-in hybrid vehicles, one with an all-electric range of 20 miles (HEV 20), and one with an all-electric range of 60 miles (HEV 60). All of the HEVs were assumed to use state-of-the-art nickel-metal hydride (NiMH) batteries and regenerative braking, and to have similar performance, including a minimum top speed of 90 mph and a 0–60 mph acceleration time of less than 9.5 seconds. All vehicles were assumed to have sufficient gasoline storage to provide a range of 350 miles. The main design features of the various vehicles are summarized in Table 9.3, which shows that the HEV 60 requires a much greater battery capacity, but has a smaller engine than the other hybrid configurations or the CV. It is interesting to see that even though the battery in the HEVs add considerable mass, the overall vehicle mass is less than that of the CV, except for the HEV 60, but even this is only 100 kg heavier

Table 9.3. *EPRI study vehicle configurations*

Vehicle	CV	HEV 0	HEV 20	HEV 60
Gasoline engine power, kW	127	67	61	38
Electric motor power, kW	–	44	51	75
Battery capacity, kWh	–	2.9	5.9	17.9
Vehicle mass, kg	1682	1618	1664	1782

than the CV. This is primarily due to the much smaller, and therefore lighter, gasoline engine and drivetrain used in the HEVs. Not surprisingly, given the similar performance expectations, the total power available from both the engine and the electric motor in the HEVs is only slightly less than that of the engine in the CV. The power split between engine and electric motor is approximately equal for the HEV 0 and the HEV 20, while for the HEV 60 the electric motor has about twice the power of the gasoline engine.

In considering the overall energy consumption of all of the vehicles the study took a "well-to-wheels" approach, and included the energy obtained from the gasoline on-board the vehicle and the energy required to process the crude oil to produce the gasoline, as well as the electrical energy required to recharge the batteries for both of the plug-in hybrid vehicles, the HEV 20 and the HEV 60. The energy required to produce the gasoline from crude oil, and the primary energy required to produce the electricity required for recharging the two plug-in hybrids was referred to as the "fuel cycle" energy. For the battery recharging part of the process, the study assumed that electricity would be generated from natural gas using a combined cycle power-plant, with an overall thermal efficiency of 50%. Many different performance parameters were calculated during the simulation, but the main results of the study can be summarized by reference to Figure 9.10, showing the CO_2 emissions per mile traveled, assuming a real-world driving schedule.

The total CO_2 emissions shown in Figure 9.10 provide a measure of the overall "well-to-wheels" efficiency of the vehicle and fueling system, as well as the overall contribution to greenhouse gas emissions. For the CV the total CO_2 emitted per mile driven includes that generated in processing the crude oil into gasoline in the fuel cycle, as well as from burning the gasoline used by the vehicle engine. This is the same situation for the stand-alone hybrid vehicle, HEV 0, since only

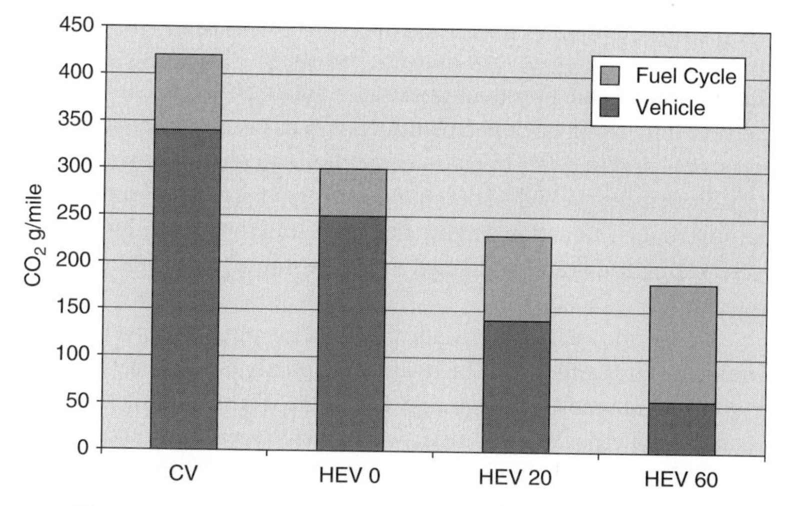

Figure 9.10 CO_2 emissions from EPRI study. *Source*: EPRI (2001). Comparing the benefits and impacts of hybrid electric vehicle options. Electric Power Research Institute Report 1000349.

gasoline is used, and the fuel-cycle energy represents the same proportion of total energy used in the CV. However, because of the much higher efficiency of the HEV 0 compared with the CV, the total emissions are reduced by nearly 30%. By moving to the HEV 20 vehicle, the total CO_2 emissions are now just over half those of the CV, while for the HEV 60 they are reduced by nearly 60%. However, it can be seen that for the HEV 60 nearly 70% of the total emissions are produced from the generation of electricity, which is assumed to use natural gas as the primary energy source. These results point to the importance of moving to a plug-in hybrid vehicle strategy in conjunction with a sustainable electrical energy supply system in the long term. If, for example, the electricity used in the overall energy conversion chain were derived from some combination of renewable energy and nuclear power, then the only CO_2 emissions would be those due to the on-board consumption of gasoline, and a small amount for fuel processing, so that the total CO_2 emissions for the HEV 60 would be about 65 g/mile. This would then mean that by moving from a vehicle fleet of all conventional vehicles, to one with all HEV 60 vehicles, together with a zero-emission electricity system, CO_2 emissions would be reduced from 420 g/mile to 65 g/mile, a reduction of 85%. The widespread use of plug-in hybrid vehicles, therefore, together with a move to a zero CO_2 emission electricity grid, would go a long way towards eliminating the greenhouse gas contribution from motor vehicles. Another very important advantage of plug-in HEVs, which would likely run mostly in an all-electric mode in city centers, would be the nearly complete elimination of smog-producing exhaust emissions in urban areas. In the very long term, once petroleum resources become very scarce and expensive, and in order to completely eliminate greenhouse gas emissions from motor vehicles, the small quantity of liquid fuel required for the internal combustion engine could be obtained from biofuels, or even from hydrogen produced from sustainable primary energy sources.

Given the very obvious advantages that plug-in hybrid vehicles would appear to offer, and the relative simplicity of using the electrical grid to provide most of the energy for HEVs, it is surprising that there has not been more development work in this area. The reduced complexity and cost of plug-in HEVs compared with hydrogen-powered fuel-cell vehicles for example, should make them much more attractive as an alternative to the existing vehicle fleet. Also, the expansion of the electrical grid required for the "fueling" infrastructure is likely to be much less expensive than developing a completely new hydrogen fueling infrastructure. Expansion and upgrading of electrical generation capacity and distribution networks is quite straightforward, and this could be phased in over time as the number of plug-in hybrid vehicles expands. The type of infrastructure required in city-center parking lots is just like that already used in some cold-weather cities, such as Edmonton, Canada, to ensure that vehicles can be started in sub-zero temperature conditions. Electrical outlets are provided at many urban parking stalls in Edmonton so that the "block heater" installed in most vehicles can be connected while the car is parked for long periods of time in temperatures down to −30 °C. If vehicles are parked overnight outside, as is often the case in suburban areas, they are also usually "plugged in" to prevent the engine coolant from freezing during very cold weather conditions. A simple expansion of this kind of infrastructure would provide all of the electricity required for recharging propulsion batteries while parked in urban centers during a typical 8-hour working day. There could even be an added benefit from such a system in providing an improved load factor for electrical utilities by spreading the electrical load out more evenly during the day. With many commuters plugging their cars in for recharging during the working day and overnight, the increased electrical load, which is normally low at these times, will ensure that electrical generation capacity is better utilized. Also, during the morning and evening rush hours, when most commuters are traveling, the recharging load will be reduced during the same period when home demand normally peaks due to the use of electrical appliances and lights at meal times. This "load-leveling" could provide a significant improvement in utility load factors, with the result being better use of generating equipment and a reduction in unit electricity generation costs.

The publication of the EPRI study has created a lot of interest in plug-in, or grid-connected, HEVs, and at least one manufacturer is now undergoing trials of a small number of vehicles. The successful commercialization of plug-in HEVs will be greatly helped by the development of better batteries, and this is being pursued by several government laboratories as well as by manufacturers. Nickel metal hydride batteries, with an energy density and specific energy of about twice that of lead-acid batteries, have now become the state-of-the-art for grid-independent HEVs like the very successful Toyota Prius. In the longer term, for use in plug-in HEVs, these may be replaced by lithium ion batteries, or with new lithium polymer or even lithium sulfur technology which is at an early stage of development, but shows considerable promise.

For local transportation in urban centers the inevitable rise in gasoline and diesel fuel prices, as well as higher parking and congestion charges used to discourage personal automobile use, should also result in mass transit being a much more popular alternative. In cities with a large population, and high population density, this will normally be in the form of guided transit such as underground subway trains, or a similar light rail transit system using an elevated guideway. For either system, electricity is the obvious choice for the propulsion energy, and if this is obtained in the long term from sustainable primary energy sources, then it will also make a major contribution to reducing greenhouse gas emissions, as well as to reducing urban pollution levels. In smaller cities, diesel-powered buses are normally the mainstay of urban transit systems, but these will not be a sustainable option in the long term. Over time these can quite readily be replaced with either trams, using a street level guideway, or with trolley buses, powered by overhead electrical catenaries, similar to those used for electrified railways. These electric trolley buses have been successfully used for many years, particularly in cities like Vancouver, Canada, which have very low electricity costs, but these should be increasingly attractive as an alternative to diesel buses or automobiles. Again, this will result in a much more sustainable urban transit infrastructure if the electricity is generated predominantly from sustainable sources, such as renewable energy or nuclear power.

9.3 TRAINS, PLANES, AND SHIPS

The "fuel" of the future for railroad transportation is also clearly electricity, and in the longer term this will likely become much more sustainable than it is currently in most countries. A truly sustainable electricity generation system, will likely use some combination of renewable energy and nuclear power, or perhaps even fossil fuels with carbon capture and storage. Electrification of railways is already well-established, and is used to serve both passenger traffic and the transport of freight in regions of the world with high traffic densities. With increasing prices for diesel fuel, and concern about global warming from greenhouse gases, there will be a clear incentive to switch from diesel-electric to "grid-connected" all-electric locomotives. It is much less clear, however, to see what a sustainable fuel supply might look like for both the marine and air transportation sectors, as these are obviously not amenable to electrification. In both cases the advantages of very high energy density, and ease of storage, offered by liquid fuels is difficult to replicate.

Given that liquid fuels will likely continue to be the only practicable choice to power ships and aircraft, it may well be that these will just continue to use liquid fossil fuels in the form of diesel fuel, heavy bunker oil, and aviation jet fuel. If the road transportation system, including rail transportation, which together account for 85% of transportation energy use, is largely converted to operate on sustainable electrical power, or even on hydrogen produced in a sustainable manner, then the energy consumption by the marine and air sectors would be quite a small fraction of overall global energy consumption. These sectors could therefore probably continue to operate on fossil-derived liquid fuels, without the need for carbon capture and storage, and make only a minor contribution to greenhouse gas emissions. If a global consensus is reached (however unlikely this may be) that there must be absolutely no greenhouse gases emitted anywhere in any energy conversion chain, then the only alternative for these sectors is likely to be biofuels or liquid hydrogen. The biofuel could be biodiesel fuel for use in marine diesels, and this could also be used in jet engines with some engineering development work. Another possible biofuel route could be methanol or ethanol for use in spark-ignition marine engines, but alcohol fuels are not very suitable for aircraft use, as the energy density is too low. The gravimetric energy density (or specific energy) of methanol, for example, is about one-half that of jet fuel, which would mean that the weight of methanol required would be twice that of jet fuel for a given energy output. For an airplane, in which the fuel load is a significant fraction of the total take-off weight, the use of methanol as a fuel would severely limit the payload.

A longer term, and probably much more expensive alternative, would be the use of hydrogen in both airplanes and ships. Although some engine changes would be needed for both marine and aircraft engines, these would be fairly minor, and in fact, the US Air Force first flew an airplane with one engine operating on hydrogen fuel in 1956. For aircraft usage the fuel would almost certainly be liquefied hydrogen, as the problems of storing it at a temperature of −250 °C, would be somewhat alleviated at the normal ambient temperature of around −60 °C at a cruising altitude of 35 000 feet. Liquid hydrogen also has a high specific energy (energy per unit mass), which would reduce the weight of the fuel required by a factor of approximately 2.5 compared with jet fuel. This reduced fuel weight for a given energy content would be an obvious advantage as an aircraft fuel, but this is overcome by the low energy density (energy per unit volume) of liquid hydrogen compared with jet fuel. The volume, and therefore the fuel tank size, required for liquid hydrogen is about four times that of jet fuel. Studies by NASA, for example, have shown that the additional space, and therefore airframe weight, required for the much larger fuel tanks would just about completely negate the advantage of the reduced fuel weight. Liquid hydrogen could also be quite easily used for ship propulsion, in a similar manner to the way in which liquid natural gas is now used to fuel LNG carriers. The additional volume required for hydrogen fuel would not be such a disadvantage for use in ships which have a much higher ratio of cargo volume to fuel volume than do aircraft.

The concept of a liquid hydrogen-fueled airplane was seriously considered over a 3-year period from 2000 to 2003 by a consortium of the Airbus company and a number of other European aerospace companies, and the resulting design was dubbed the "cryoplane" due to the cryogenic storage tanks which would be used to store liquid hydrogen at a temperature of −250 °C. Several different configurations were studied, ranging from a small business size jet up to a large passenger airplane of similar size to the Airbus A380. The study participants found that the concept of a hydrogen-fueled airplane was feasible, and that jet engines could be modified to run on liquid hydrogen fuel with no decrease in efficiency. Different storage tank locations were considered for the different size airplanes studied, and a long-range airliner configuration with cryogenic fuel tanks located in the expanded upper fuselage portion of the airframe was found to be a likely option.

BIBLIOGRAPHY

EPRI (2001). *Comparing the Benefits and Impacts of Hybrid Electric Vehicle Options.* Electric Power Research Institute Report 1000349.
Rovera, G. (2001). *Potential and Limitations of Fuel Cell in Comparison with Internal Combustion Powertrains.* Fiat Research. Presented at *ICE 2001*, Capri, Italy.
Toyota Motor Corporation (2005). *http://www.toyota.com/*
US Department of Energy (2005). *http://www.energy.gov/*
US Department of Energy. Argonne National Laboratory (2005). *http://www.anl.gov/*
World Energy Council (2005). *http://www.worldenergy.org/wec-geis/default.asp*

10

Achieving a sustainable energy balance

In the preceding chapters we examined current energy demand and supply patterns, as well as some projections for future global energy demand. We then discussed the fact that all of our energy requirements must be met, ultimately, from some combination of only three primary energy sources; fossil fuels, renewable energy, and nuclear power. Increasing concerns about the environmental effects of large-scale fossil fuel usage, as well as uncertainties about the long-term availability of these fuels, particularly crude oil, provided the background for further exploration of alternative energy supply strategies. We discussed the need to move, in the long term, from an overwhelming reliance on fossil fuels to provide nearly 80% of our total energy requirements, to a more sustainable energy supply mix. We went on to discuss the prospects for some of the alternatives to fossil fuels, including increased use of renewable energy and nuclear power, primarily to generate electricity. Consideration was also given to a more sustainable way of using fossil fuels by capturing and storing carbon dioxide, although the technology for doing so is at an early stage of development. Finally, we speculated that the move away from fossil fuels may, in large part, come about as a result of changing from petroleum fuels to electricity as the energy carrier to supply most of our transportation energy needs. In this final chapter we will examine how all of these ideas may come together, and speculate on how the primary energy supply mix could evolve over the remainder of the twenty-first century. Changes in this mix over time will result not only from technological improvements and reductions in the cost of these energy alternatives, but also from the development of rational energy and environmental policies around the globe.

The goal of any energy policy, whether it is regional, national, or multinational in scope, is to provide a reasonable balance between energy demand and energy supply in all economic sectors. The word "reasonable," of course, may be interpreted quite differently by different people. To the consumer, struggling to pay ever-escalating energy bills, reasonable might be interpreted to be "reasonable cost," or at least prices that aren't escalating by more than the cost of living. To the ardent environmentalist, reasonable might be interpreted to be nothing less than an almost total reliance on renewable energy to satisfy all of our energy needs. And, finally, to the chairman of a global energy company reasonable might be interpreted to be prices that enable his company to earn attractive profits while still spending the very large sums of money needed to find new hydrocarbon resources, or develop new technology aimed at reducing our dependence on fossil fuels. The end result, as in most matters of public policy, is usually a compromise between the many different factors that affect the supply of energy. Government and corporate leaders, however, are increasingly striving to develop policies that will result in a steady supply of energy at affordable prices while at the same time minimizing the effects of energy use on our environment.

To study the energy demand–supply balance in more detail, we will return to the energy flow diagram, or "Sankey" diagram, which we introduced in Chapter 2. This very informative diagram can be used to track how energy is converted and distributed, from primary source all the way through to end-use. The Sankey diagram can be used to visualize these complex flows of energy, either for a single nation or national region, or even for total global energy flows. They can also be very useful in providing a quick visual snapshot of the quantity of primary energy that becomes "useful energy" in supplying various end-uses, and the quantity of "unavailable" energy that ends up being rejected, usually in the form of waste heat. An illustration of this type of diagram for the complete US economy, which has been prepared by the Lawrence Livermore Laboratory of the US Department of Energy, is shown in Figure 10.1 (US DOE, 2005). Sankey diagrams are available from the DOE for several years, but the one shown in Figure 10.1 for 2002 is particularly useful because the total amount of primary energy consumed, including that consumed for non-energy uses, such as the petroleum used for chemical feedstock and asphalt production, just happens to total 103 EJ (Exajoules, or 10^{18} joules). This then makes it very easy to determine the approximate percentage flows of energy directed to various end-uses, as well as to waste heat or "rejected energy." For example, the diagram clearly shows the various flows of

U.S. Energy Flow Trends – 2002
Net Primary Resource Consumption ~103 Exajoules

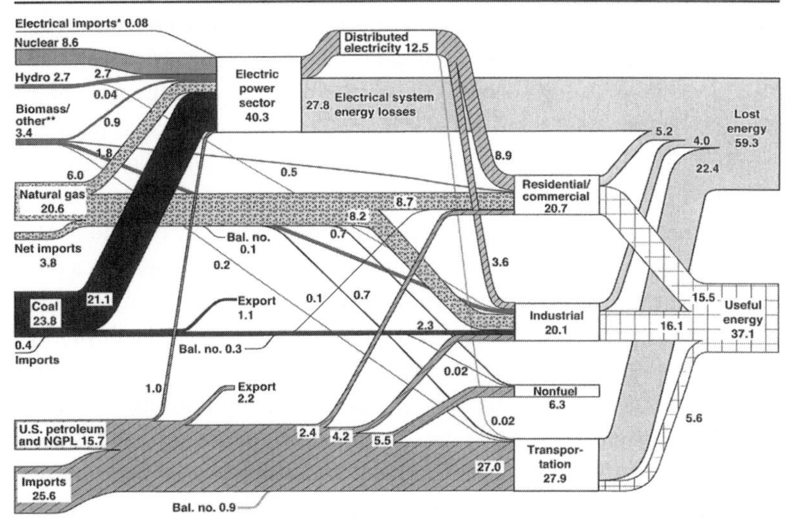

Figure 10.1 Energy flow diagram for the USA - 2002. Energy flow totals ~ 103 EJ. *Source*: Derived from US Department of Energy, Energy Information Administration, *Annual Energy Review 2002*, DOE/EIA-0384(2002). Washington, DC, October 2003.

primary energy being used to generate electricity, with the largest source being coal, accounting for 21.1 EJ, followed by 8.6 EJ of primary energy input in the form of nuclear energy, 6.0 EJ from natural gas, and 2.7 EJ from hydro power. However, of the total of 40.3 EJ used for electricity generation, only 12.5 EJ ends up as electricity, while 27.8 EJ ends up as rejected energy, primarily in the form of waste heat from thermal power stations. Similarly, it can be seen that in the transportation sector, which relies primarily on petroleum as a source for the 27.9 EJ of primary energy used, just 20% or some 5.6 EJ, is turned into "useful energy" to propel all the cars, trucks, ships, and airplanes. Also, 6.3 EJ of petroleum supplies are used for "non-fuel" applications, such as the production of plastics, and other industrial materials. In total, then, of the 103 EJ of all forms of primary energy used in the USA in 2002, only 37.1 EJ was "useful energy" to provide heat and power to homes, factories, and vehicles, and 6.3 EJ was used in non-fuel applications, while some 59.3 EJ was "lost energy," or unavailable energy, primarily in the form of waste heat.

Figure 10.1 shows that renewable energy provides only a small percentage of total primary energy requirements in the USA, with the largest component being the 2.7% of total energy derived from hydroelectric power. Another 3.4% of primary energy is derived from biomass and "other" renewable energy forms, with the largest fraction being biomass energy which is used in industrial processes, such as pulp and paper mills. In 2002 only some 0.9% of total primary energy was obtained from renewable sources other than hydroelectric power to generate electricity. The figure also illustrates the heavy dependence of most Western nations on petroleum, or crude oil, primarily as a feedstock to supply transportation energy needs, and on coal, which is used mainly to generate electricity. In the USA, as in most of the developed world, much of the primary energy in the form of crude oil is imported, leading to concerns about energy security. Long-standing concerns about energy security, as well as more recent concerns about greenhouse gas emissions, is causing most nations to seriously examine alternatives to their heavy reliance on fossil fuels as the predominant form of primary energy. From Figure 10.1 it can be seen that in 2002 nearly 90% of all primary energy in the USA was derived from fossil fuels in the form of coal, crude oil, or natural gas, all of which result in the production of carbon dioxide, the most important source of greenhouse gas emissions. In the remaining part of this chapter we will look back at some of the alternatives to fossil fuel use that we have discussed in order to examine how we may begin the transition to a much more sustainable energy demand–supply balance.

A move away from fossil fuels to supply most of our primary energy needs means relying more heavily on renewable energy and/or nuclear power wherever possible. One way to achieve this, as we have discussed in Chapter 9, is to reduce our almost total reliance on crude oil to provide transportation energy by moving to electricity rather than refined petroleum products as the main transportation energy carrier. This would then mark the beginning of a transition from our present-day "hydrocarbon economy" to a new "electricity economy." As we have seen in Chapter 4, transportation accounts for more that 25% of total primary energy consumption, so a large-scale transition to the plug-in hybrid vehicles we discussed in Chapter 9 would require a major expansion in electricity production. If this expansion was to be mainly from fossil fuels, without the use of carbon sequestration, then there would be little reduction in the production of greenhouse gases. The development of a truly sustainable electricity supply, capable of satisfying the energy needs of our transportation sector, as well as all

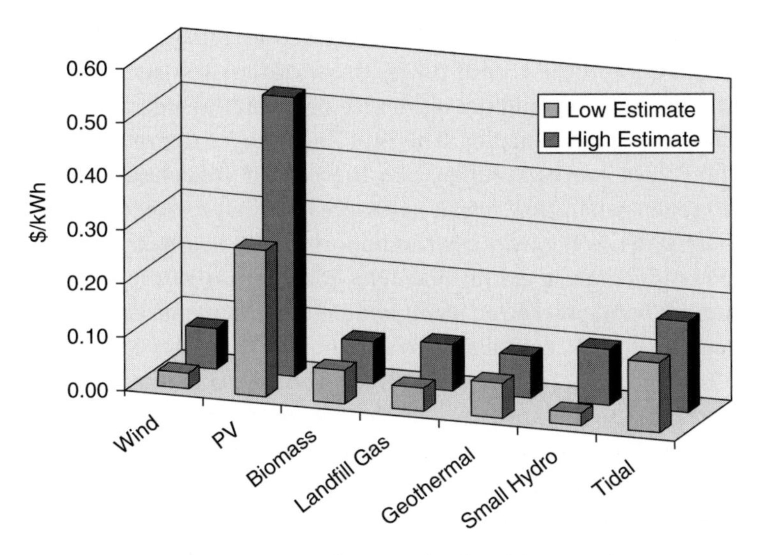

Figure 10.2 EU estimates of renewable electricity costs in 2005.
Source: EU Atlas Project.

other economic sectors, would require a transition to some combination of renewable energy, nuclear power, and clean coal with carbon sequestration as primary sources. The actual mix of these sources that develops over the next 50 to 100 years will depend on their technical development, cost, and public acceptance, all of which will inform and influence future energy policies. We have seen, for example, that renewable energy is increasingly being used to generate electricity, but that the low energy density of most forms of renewable energy has resulted in high costs and significant impacts on land-use. On-going technical developments have contributed to reducing the costs of some forms of renewable electricity generation, however, with wind-power leading the way into the mainstream as an important source of electricity.

A study of the relative costs of many forms of electricity generation from renewable energy was conducted under the European Union's "Atlas" project in 1996 (European Atlas Project, 2005). The estimated range of renewable electricity costs in 2005 is shown in Figure 10.2 for the major technologies studied. Both a high estimate and low estimate of the cost of electricity (converted from euros to $/kWh) is shown, and the range between each varies for each technology studied. Some technologies, such as wind power, solar photovoltaic power generation (PV), and small hydro generation, are particularly sensitive to the site chosen and the range between low and high cost estimates is large in these cases. In other cases, like biomass or geothermal energy, the cost of power production is less sensitive to the particular location, and the difference between the high and low cost estimates is much smaller. The unit electricity costs of each of the technologies can be compared to the cost of conventional PF coal-fired power generation, which in Chapter 8 we have seen to be approximately $0.042/kWh, without the imposition of a carbon tax or other form of greenhouse gas disincentive. From Figure 10.2 it can be seen that both wind power and small hydroelectric generation can be very competitive with coal-fired power generation at the low range of the cost estimates. Also, under certain conditions and in particularly favourable sites, it appears that the use of land-fill gas, biomass, or geothermal energy for electricity generation can approach the costs of fossil-fuel generation. On the other hand, solar photovoltaic power generation, although benefiting from important reductions in capital costs, and tidal power are uncompetitive with conventional coal-fired power generation at this time. It would appear, therefore, that there is considerable scope for greater penetration of renewable primary energy for electricity generation, particularly if the benefit of eliminating greenhouse gas emissions is taken into account. As this penetration increases, however, the growth of renewable electricity may be hindered by growing public opposition associated with the large land areas required by low energy density sources such as wind power. This is already starting to be an issue in Europe, for example, where protests from local countryside groups greet many new proposals for the development of large wind farms. These protests are increasingly gaining the attention of the general public and politicians in the UK, and in Germany, the largest producer of wind power in the world, where countryside campaigners claim that large wind farms are destroying vast areas of natural beauty. These types of protests, together with the intermittent nature of many renewable energy sources, as we discussed in Chapter 7, will likely limit its penetration into the electricity generation mix.

As we have seen in Chapter 8, there is also increasing interest by electric utilities, and government, in nuclear power as a clean source of electricity. There is a growing realization that nuclear power may be the most important way to reduce our impact on the global climate, and public acceptance of nuclear power also appears to be improving. Many countries may look to the example of France, in which most of the electricity is generated by nuclear powerplants with very little opposition from the general public. One scenario we may envisage

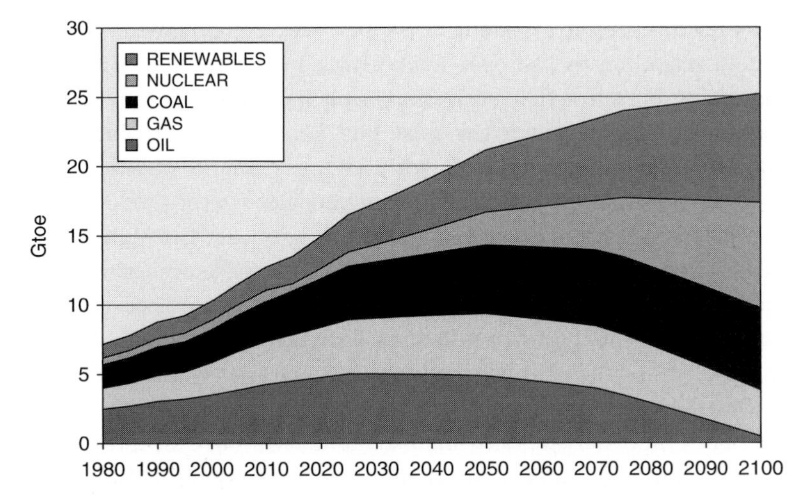

Figure 10.3 World primary energy supply – Nuclear and renewable energy scenario.

for the future, therefore, is a long-term transition from refined petroleum products to electricity as the dominant energy carrier. Moving away from relying predominantly on hydrocarbons as our main primary energy source, and towards a much greater utilization of renewable energy and nuclear power to generate electricity, could lead to a much more sustainable energy future. The effects of this will likely be seen most clearly in the transportation sector, where electricity may replace liquid hydrocarbon fuels as the energy carrier of choice, as we have discussed in Chapter 9. These ideas can be used to speculate on how the primary energy mix may change over the next 100 years, as illustrated for a possible "Nuclear and renewable energy scenario" in Figure 10.3. This figure demonstrates the type of projections which may be made by simply assuming plausible growth rates for each of the three primary resources; fossil fuels, renewable energy, and nuclear power. The projected primary energy demand, in billions of tonnes of oil equivalent (Gtoe), is shown for the remainder of the twenty-first century, with actual data shown for the last 20 years of the twentieth century. It should be emphasized that the projected primary energy mix shown in Figure 10.3 is not the result of complex (or even simple!) economic modeling of the economy, but is just an attempt to show the effects of varying the take-up rates for each primary resource.

The first thing we need to do to construct a chart like the one shown in Figure 10.3 is to assume a growth rate for overall world energy demand. As we have seen in Chapter 4, the overall world energy demand has grown by about 1.75% per year compounded over the last 20 years, but by just over 4% in China and 6% in India. Given the increasing presence that both China and India are likely to have in the world economy over the next few decades, the projections in Figure 10.3 assume an overall world energy demand growth rate of 2% per year from 2000 to 2025, which is very close to the predictions of the IEA for this period, as we have seen in Figure 4.6. The high growth rates are unlikely to be sustained forever in these emerging economies, however, and as energy costs continue to escalate we have assumed an annual growth rate of 1% per year from 2025 to 2050. For the last half of the century, we have assumed that world population growth will be decreased and that increasing emphasis on energy efficiency and "demand side management" will have an impact. We have therefore assumed a world energy growth rate of 0.2% per year from 2050 to 2075, and finally 0.1% from 2075 to the end of the century. Even with these modest growth rate assumptions the total world consumption of primary energy would grow by a factor of 2.5 from some 10 Gtoe in 2000 to 25 Gtoe in 2100. To construct Figure 10.3 we speculated that the use of crude oil would decline dramatically towards the end of the century, due to declining resources and increased costs as we discussed in Chapter 5. Dwindling supplies of petroleum near the end of the century would likely be used to provide the "back-up" fuel required for the internal combustion engines in plug-in hybrid electric vehicles, although eventually this may be replaced by bio-fuels. We have also assumed a similar, but smaller, decline in the use of natural gas due to its relatively better availability, and an assumption of increased global trading of "locked-in" gas using new pipeline capacity and greater use of Liquefied Natural Gas (LNG).

With the assumptions used to construct Figure 10.3, the share of fossil fuel use as a fraction of total primary energy demand has dropped dramatically, from 80% of total demand in 2000 to just 39% in 2100. The actual consumption of fossil fuels has actually increased slightly, however, due to the increased consumption of "clean coal" assumed for electrical power generation. The share of nuclear power use over the century has increased from 6.8%, however, to over 30% of total primary energy, while renewable energy has increased from 13.6% to 31% of the primary energy supply. The large increase in the use of nuclear power and renewable energy towards the end of the century, together with most of the coal, would be used to generate electricity, thereby speeding the transition to an "electricity economy." Much of this new supply of electricity would be used to supply the transportation sector, and some would be used to significantly increase the share of electricity used for space heating, mainly through the widespread adoption of heat pumps. Some may question the feasibility of this expansion, given the significant increase in electricity infrastructure that such a transition would imply. This type of expansion has been done in the past, however, first of all around the beginning of the twentieth century when electricity was a relatively new energy carrier, and was being used to replace gas lighting and steam engines. A second major expansion in the electricity infrastructure took place in the mid-twentieth century, particularly in North America with widespread rural electrification programs. Given the relatively short time-frame required for these earlier transitions, a further expansion and move towards the electricity economy should certainly be feasible over the next 100 years.

If, for some reason, the use of nuclear power and renewable energy does not expand to the extent assumed in the Nuclear and Renewable Energy Scenario, another alternative for enabling the transition to an electricity economy is the "Clean Coal Scenario." In this scenario, coal use would be greatly expanded, and would be used primarily to generate electricity, and perhaps also to produce synthetic liquid and gaseous fuels. In order to reduce greenhouse gas emissions, the use of "clean coal" technology, together with carbon sequestration would need to be widely implemented. New coal-fired powerplants would likely be based on the IGCC approach described in Chapter 6, leading to increased efficiency and reduced emissions. Additionally, the carbon dioxide generated by such plants would need to be "sequestered," using carbon capture and storage techniques which are still in the earliest stages of development. This aspect of clean coal technology is probably the least well developed at the present time, and much more work needs to be done to determine if it will be technically and economically viable on the very large scale required to sequester most of the CO_2 produced. Although preliminary trials of carbon sequestration have been undertaken, as discussed in Chapter 6, much more work needs to be done to determine if there really are enough suitable repositories for the long-term sequestration of the huge volumes of carbon dioxide that would be released by the combustion or gasification of the large quantities of coal that would be used in any clean coal scenario. Nevertheless, a possible primary energy mix for such a strategy is shown in Figure 10.4, using the same energy demand assumptions as in Figure 10.3. In this scenario we have also assumed that both nuclear power and renewable energy would still play a significant role

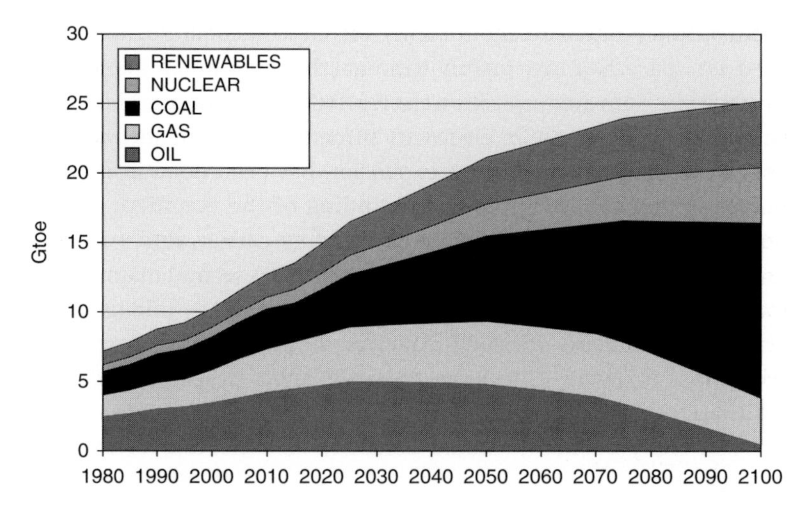

Figure 10.4 World primary energy supply – clean coal scenario.

in the primary energy mix, but with much smaller annual growth rates compared with the previous scenario. We have also made the same assumptions about the declining availability of crude oil and natural gas as in the renewable energy and nuclear power scenario. The share of total primary energy supply provided by coal under this new scenario has then been assumed to increase in every decade, reaching 50% of total primary energy supply by the end of the century compared with 23% in 2000.

Of course, the actual mix of primary energy sources that will develop over the remainder of this century is likely to be somewhere between these two scenarios. The primary energy supply mix that evolves over time will depend on both advances in technology and on the priority which individuals and governments give to developing cleaner and more sustainable energy sources. In this book we have focused primarily on technological solutions to developing a more sustainable energy supply, but we should not forget that one of the most important options open to mankind is to simply reduce our demand for energy in the first place. This is sometimes referred to as the "soft-side" of energy policy, but programs aimed at convincing corporations, governments, and individuals to use energy more efficiently, and to simply avoid wasteful or frivolous use, will be powerful tools in the quest for a more sustainable economy. Success with these types of programs, whether they are primarily aimed at "energy conservation" on the part of individual users, or more sophisticated "demand-side management" policies for large corporations and utilities, requires widespread behaviour modification. Policymakers need to be aware, therefore, that the development of sustainable energy policies needs the wholehearted participation not only of scientists, engineers, and economists, but also of social-scientists and the general public at large if they are to be successful. What appears quite clear, however, is that there are viable solutions to the quest for cleaner energy supplies which should be sufficient to provide all of our requirements for the foreseeable future. It's now up to all of us: corporate leaders, politicians, and individual consumers, to play our part in seeing that our energy future is a truly sustainable one.

BIBLIOGRAPHY

European Atlas Project (2005). *http://europa.eu.int/comm/energy_transport/atlas/homeu.html*

US Department of Energy (2005). *http://www.energy.gov/*

APPENDIX: ENERGY CONVERSION FACTORS

Energy conversion factors:					
1 Btu =	1055 J =	1.055 kJ			
1 MMBtu =	1 × 10*6 Btu				
1 kJ =	0.9478 Btu				
1 GJ =	0.9478 MMBtu				
1 kWh =	3600 kJ =	3.6 MJ			
1 kWh =	3412 Btu				
1 toe =	41.87 GJ =	39.72 MMBtu =	7.35 Bbls		
1 Mtoe =	41.87 PJ =	41.87 × 10*15 J =	41.87 × 10*12 kJ =	39.7 × 10*12 Btu	
1 Gtoe =	41.87 EJ =	41.87 × 10*18 J =	41.87 × 10*15 kJ =	39.7 × 10*15 Btu =	39.7 Quads
1 × 10*15 Btu =	0.025189 Gtoe				
1 Bbl =	42 US gals. =	35 Imp. gals.			
1 m^3oe =	6.292 Bbls =	0.856 toe =	35.843 GJ		
Note: toe = tonne of oil equivalent					
Prefixes:					
10*3 = kilo	10*6 = Mega	10*9 = Giga	10*12 = Tera	10*15 = Peta	10*18 = Exa
Thousand	Million	Billion	Trillion	Quadrillion	

Index